T0323534

EEG BRAIN SIGNAL CLASSIFICATION FOR EPILEPTIC SEIZURE DISORDER DETECTION

EEG BRAIN SIGNAL CLASSIFICATION FOR EPILEPTIC SEIZURE DISORDER DETECTION

SANDEEP KUMAR SATAPATHY

SATCHIDANANDA DEHURI

ALOK KUMAR JAGADEV

SHRUTI MISHRA

ACADEMIC PRESS

An imprint of Elsevier

Academic Press is an imprint of Elsevier
125 London Wall, London EC2Y 5AS, United Kingdom
525 B Street, Suite 1650, San Diego, CA 92101, United States
50 Hampshire Street, 5th Floor, Cambridge, MA 02139, United States
The Boulevard, Langford Lane, Kidlington, Oxford OX5 1GB, United Kingdom

Notices

Knowledge and best practice in this field are constantly changing. As new research and experience
broaden our understanding, changes in research methods, professional practices, or medical treatment
may become necessary.

Practitioners and researchers must always rely on their own experience and knowledge in evaluating and
using any information, methods, compounds, or experiments described herein. In using such information
or methods they should be mindful of their own safety and the safety of others, including parties for
whom they have a professional responsibility.

To the fullest extent of the law, neither the Publisher nor the authors, contributors, or editors, assume any
liability for any injury and/or damage to persons or property as a matter of products liability, negligence or
otherwise, or from any use or operation of any methods, products, instructions, or ideas contained in the
material herein.

Library of Congress Cataloging-in-Publication Data
A catalog record for this book is available from the Library of Congress

British Library Cataloguing-in-Publication Data
A catalogue record for this book is available from the British Library

ISBN 978-0-12-817426-5

For information on all Academic Press publications
visit our website at https://www.elsevier.com/books-and-journals

Working together
to grow libraries in
developing countries

www.elsevier.com • www.bookaid.org

Publisher: Stacy Masucci
Acquisition Editor: Rafael E. Teixeira
Editorial Project Manager: Samuel Young
Production Project Manager: Maria Bernard
Cover Designer: Mark Rogers

Typeset by SPi Global, India

CONTENTS

PREFACE

EEG signals are the electric waves generated in the human brain which is the main cause of different tasks performed by a person. These signals are very small in magnitude but collectively it defines the behavior of a person. Hence, careful and complete analysis of these signals can solve different diseases occurring in our brain. High volume and uncertainty of these signals make it very difficult for any medical person to diagnose a brain disorder disease just by looking at the graphical representation of these signals. Hence, there are plentiful opportunities for computer scientists to produce a computer-based model that can detect efficiently any specific neurological disease after proper analysis of these signals. The signals recorded from human brain contain several features, those needs to be extracted by some feature extraction technique. After this any disease detection problem can be solved by the application of different data mining techniques. The classification technique is basically used for solving these kinds of problems that takes the help of different supervised or unsupervised machine learning techniques.

The primary objective of this research is to classify the EEG brain signal for detection of epileptic seizures using machine learning techniques. Machine learning techniques have recently developed most efficient techniques for solving different kinds of problems. We are mainly concentrating on supervised machine learning techniques. These techniques require a prior knowledge about data that make the machine learn for recognizing new data and perform classification operation. But it is necessary to preprocess the raw EEG data collected from patients to bring them into a form feature sample-based dataset. The signals are analyzed by decomposition using Discrete Wavelet Transform (DWT) with Daubechies wavelet of order 2 up to level 4. After this decomposition different statistical features for different coefficient are collected together to build the dataset. We performed an empirical survey of different machine learning techniques applied for designing classifier model for epilepsy identification. The different techniques used are Multilayer Perceptron Neural Network (MLPNN) with different learning algorithms (such as back-propagation, resilient-propagation, and Manhattan update rule), different variations of neural network such as Probabilistic Neural Network (PNN), Recurrent Neural Network (RNN), Radial Basis Function Neural Network (RBFNN), and so on, and Support

Vector Machines (SVM) with different kernel functions (such as linear, polynomial, and RBF kernel).

Further, we have enhanced the performance of RBFNN network by implementing different swarm intelligence-based optimization techniques as training algorithm. Swarm intelligence is one of the most efficient fields for the optimization problems. The swarm intelligence methods used in this work are Particle Swarm Optimization (PSO) and Artificial Bee Colony (ABC) algorithms. This work proposes a novel hybrid technique to train RBFNN by optimizing the parameters of this network using the improved PSO algorithm and ABC algorithm. PSO is one of the most effective optimization technique used so far. But again we have modified the existing PSO algorithm to enhance the performance by adding a new technique for finding the value of the inertia weight used in the calculation of velocity update. The experimental evaluations have proved that the performance of the proposed technique is higher than RBFNN trained with gradient decent approach as well as RBFNN trained with the conventional PSO algorithm. We have also suggested a change to existing ABC algorithm, where we have used tournament selection in place of roulette wheel selection in onlookers bees phase. Besides, the operation of the proposed technique was shown to be more efficient by different experimental evaluations. For this work, ABC algorithm was chosen as more efficient algorithm because of the less number of dependable parameters and improved accuracy of RBFNN as compared to PSO algorithm.

Along with classification accuracy, specificity, sensitivity, and many other different parameters were considered for measuring the performance of classification algorithms. Also, the experimental evaluations are supported by a k-fold cross validation technique to validate the results obtained by different experiments.

CHAPTER 1

Introduction

Electroencephalogram or electroencephalography (EEG) is a trial performed on mental capacity to record the electrical activity in brain. The neural structure of the brain consists of several neurons in terms of lacs. These neurons communicate by colliding among themselves and passing information to each other. This collision leads to the generation of a very small amount of electricity. This is utterly different from the general electricity, which is very high in magnitude. This electric signal flow decides the behavior of a person. In a human brain the normal stream of electrical signal leads to a healthy person. And an abnormal electrical signal flow can pass to an unhealthy person. Hence, these signs can be recorded and analyzed to solve many neurological disorder diseases. The transcription of the electrical activity is essentially caused by putting electrodes on the scalp for 20–40 min, which evaluates the potential fluctuations in the brain [1]. The nerve cells in the psyche are the origin of electric charge, and then they exchange ions with the extracellular milieu. Ions of the same charge repel each other and in this manner they are forced out of the neurons when a number of ions are driven out at the same time they promote each other and form a way known as volume conduction [2]. When this wave reaches the electrode they push or force the ions along the air foil of the electrode which create potential difference and this voltage difference recorded over time gives EEG signals. The key motivation behind this research work is the rapid growth in volume of biological and clinical data or records. To extract knowledge from these data which can be served to be a clinical application, there are different data analysis difficulties which need to be overcome. Many analytical tools based on machine learning (ML) approaches have been invented to tackle with such challenging task of data analysis problems. Around 1% of the total population in the world are affected by a neurological disease called epilepsy. A careful analysis of these EEG signals can solve many neurological disorder diseases.

EEG Brain Signal Classification for Epileptic Seizure Disorder Detection
https://doi.org/10.1016/B978-0-12-817426-5.00001-6

1.1 PROBLEM STATEMENT

Nowadays, the recording of EEG signal can be easily managed with the aid of various hardware and software techniques. By simply seeing at these signals with naked eyes one cannot make out any abnormality in the sign. Hence, the most important problem is to study these signals properly and extract the hidden features present inside. A neurological disease can occur in a human brain due to abnormal EEG flow. This abnormality should be properly analyzed to specify the pattern of this disease that can help with prediction of any such type of diseases in human brains. EEG recording generally leads to the collection of a huge amount of numerical information that consists of the state of electrical activity at different time. This recording is generally taken for 10–15 transactions. This duration is sufficient to understand the state of a human brain which leads to collection of huge quantities of data. By plotting this information graphically, we can conclude some behavior of the brain, though not completely. As a result, it is more significant to collect these data and pull information from this. In this research study, the main concentration is on the neurological disorder disease called epilepsy. The whole problem of this research work has been broadly classified into two groups. First, is feature extraction and analysis of a very nonstationary signal like EEG signal. Second, is the classification of EEG signals to detect epileptic seizures.

First, we have developed a well-defined and well-structured process for extracting the hidden features from a very transient and nonstationary signal like EEG signal. For this a signal transformation technique called as discrete wavelet transform (DWT) was used. To compare the significance of these extracted features, other features based on some mathematical computations were also extracted.

Second, we have developed a well-defined and most efficient classifier model that can identify and distinguish epileptic seizures from nonepileptic ones. For this, we have considered mostly *ML-based classification techniques* like ANN based classifiers, support vector machines (SVM), and evolutionary theory-based classifiers.

1.2 GENERAL AND SPECIFIC GOALS

A very careful analysis of EEG signal can provide response for many neurological disorder diseases. These signals are mainly responsible for different abnormalities generated in the human brain. Decades ago it was not possible to study these signals due to high price of the equipment and nonavailability of adequate engineering sciences. But nowadays these problems have been resolved and many hardware equipment are available with a reasonable price for recording of the EEG signal. But still technological improvement in the depth psychology of these signals is getting along. Many researchers are putting their enormous efforts in this area.

Recording of EEG signal is generally made for approximately 20–30 min. In that respect several electrodes are placed on the human scalp to record these EEG signals. Normally a plenty of data are brought forth during this recording, which is not possible for a medical person to analyze them in naked eyes. Hence, there are several technologies developed to record these signals and directly plot the graphs by using software, which are viewed by a specialist and some observations can be made on the state of the brain. Routine EEG is a process which is applied for diagnosis in the following conditions.

1. To distinguish between epileptic seizures from nonepileptic seizures, syncope (fainting), subcortical movement disorders, and migraine variants.
2. To serve as an adjunct test of mind destruction.
3. To prognosticate in certain situations like patients with coma.
4. To determine whether to wean antiepileptic medications.

In certain situations the routine EEG is not sufficient, eventually the patients have seizure. In this case, the patients are taken to hospital and EEG is constantly recorded. This is mostly done to distinguish between epileptic seizures and other seizures. The most important advantage of EEG is its speed. A very complex neural activity can be read within a few minutes or in terms of seconds. EEG produces a very little spatial resolution as compared to magnetic resonance imaging (MRI) and positron emission tomography (PET). Thus sometimes EEG images are combined with MRI scans. According to Bickford et al. [3] the different clinical application of EEG in human and animals is used in following situations:

1. Monitor alertness, coma, and brain death.
2. Locate regions of impairment following head trauma, stroke, or tumor.
3. Monitor cognitive engagement.
4. Produce biofeedback situations.
5. Control anesthesia depth.
6. Investigate epilepsy and locate the seizure origin.
7. Test epilepsy drug effects.
8. Assist in experimental cortical excision of epileptic foci.

9. Investigate sleep disorder and physiology.

10. Monitoring amobarbital effect during Wada test.

Other than clinical uses there are different applications where EEGs are extensively used. Brain computer interface (BCI) is a typical communication system that recognizes user's command only through their brain waves and responds accordingly. It is frequently known as mind machine interface or brain machine interface. It can be used for people who are unable to express through speech or physical activities. The basic function of these devices is to intercept electrical signals that occur between nerve cells in the brain and render them into a signal that can be read by an external device. The different cases of BCI are invasive BCI, partially invasive BCI, and noninvasive BCI.

In this research work, the boundary has been limited only to the analysis and classification of EEG signals for epileptic seizure identification. Approximately 1% of the entire population is touched on by the disease known as epilepsy. The human brain is the most critical part of body that controls the coordination of human muscles. The transient and unexpected electrical interferences in the brain result in an acute disease called as epilepsy or epileptic seizure. It is one of the most common diseases around the globe, striking more than 40 million people worldwide.

1.3 BASIC CONCEPTS OF EEG SIGNAL

The EEG signal is normally employed for the purpose of recording down the electrical actions of the brain signal that typically originates in the human brain. An EEG signal is measured in terms of currents that flow during excitation of synaptic activities of many neurons in the cerebral cortex [4]. During this excitation of brain cells, the small magnitude of currents are produced within the dendrites. As a result, these currents generate a magnetic field which can be measured by electromyogram machines.

Depth psychology and measurement of EEG signal is applied for testing the state and activity of human brain. Like PET and MRI, EEG method is also used widely due to its power of providing best temporal resolution and low price. Human brain consists of various neurons. When there will be any flow of information into the brain, these neurons hit each other. During this procedure electricity is generated, which are very small in amount and transient in nature. This is likewise called as a nonstationary signal, because the frequency of this signal is not set up for a certain amount of time. Biologically, it can be assumed that, during the activation of brain cells, the synaptic currents are created within the dendrites [4].

EEG signals were first measured by Hans Berger in 1929. He was the first person to record electrical activities in the human brain. These signals are generated by collision of various neurons and can be read by putting electrodes on the human scalp. It gives a common view of neural activity and can be employed for the field of the physiology of the human brain. These signs are very low in magnitude and therefore calculated in micro volt. These signals change according to the state of mind. Different behavior of human being can give rise to different frequencies of EEG signals. EEG is the test performed on the human brain to record these signals and visualize the state of a brain. A very careful analysis of these signals can solve many neurological disorder diseases. More often than not, the neurons get charged by pumping of ions across their membrane. These ions are exchanged constantly by the neurons. Some charged ions repel each other. In this way, several ions repel each other through their neighbors. This process is called as *volume conduction*. Utilizing this process data flows in a mental capacity which leads to generation of small electric arc.

Brain activity is followed by using several existing methods such as MRI, magneto encephalography, functional magnetic resonance imaging [5,6], and electroencephalogram. Since the EEG has a rapid response time and is inexpensive relative to other methods, it is widely used to monitor brain activity in BCI research [4].

The general structure of a human neuron is shown in Fig. 1.1. The *soma* is the cell body of neuron and contains nucleus. The *dendrites* are extended from the soma and receive chemical message from other neurons. The *axon* transmits electrochemical signals to other neurons. The *myelin sheath* contains fatty tissue cells that insulate the electrical current running through

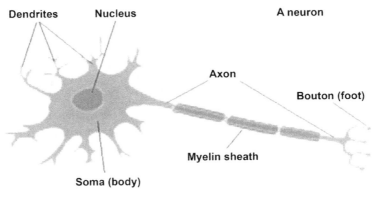

Fig. 1.1 Basic structure of a neuron.

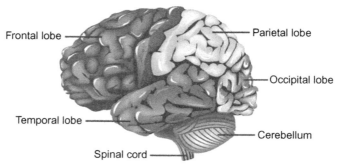

Fig. 1.2 Simple structure of brain.

the axon. The *bouton* is responsible for converting an electrical signal to a chemical signal to be received by other neurons.

The basic construction of a human brain (cerebrum) is made of two cerebral hemispheres, right and left. Each cerebral hemisphere is formed of four lobes: frontal lobe, parietal lobe, temporal lobe, and occipital lobe. These are shown in Fig. 1.2.

Cerebrum: The cerebrum makes up the greatest portion of the human brain. It is as well recognized as the cortex and it is responsible for executing several important brain functions, including action and thought process. The cerebrum is further subdivided into four different sections that bear their own respective functions and are termed as lobes.

Frontal lobe: The frontal lobe has the responsibility of executing different functions like language expressions, logical thinking, higher level cognition, and so on. Its position is at the front part of the head. Any damage to it can lead to changes of socialization, fear, and so on.

Parietal lobe: The parietal lobe is responsible for processing the data transmitted to the mind by the tactile senses like pain, pressure, and contact. It is present at the middle of the head. Any damage to it can cause problem with language, ability of controlling eye gaze, and verbal memory.

Occipital lobe: The task of occipital lobe includes the transformation of the information that is being sent to the brain by the eyes. It is present at the rear of our brain. Its damage can cause your visual ability to be affected, like unable to recognize colors, words, and so on.

Temporal lobe: The temporal lobe is responsible for making memories and processing the sounds being recorded by the ears. It is situated at the back of the brain. Any damage to it can cause problem with speech communication sciences, speech perception, and computer storage.

The recording of the electrical activity is basically done by placing electrodes on the scalp for 20–40 min, which measures the voltage fluctuations in

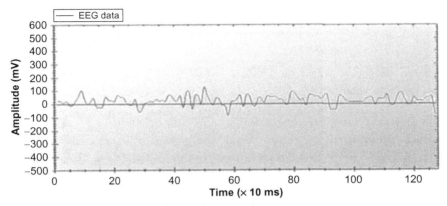

Fig. 1.3 The raw EEG record of a baseline sample.

the brain [7]. The neurons in the brain are the source of electric charge, so they exchange ions with the extracellular milieu. Ions of the same charge repel each other and in this manner they are forced out of the neurons when a number of ions are driven out at the same time they promote each other and form a way called as volume conduction [8]. When this wave reaches the electrode they push or force the ions along the air foil of the electrode which create potential difference and this voltage difference recorded over time gives EEG signals. A sample EEG recording is shown in Fig. 1.3.

Human scalp EEG was born in 1920 by the German physician that measured the brain activities on the scalp and since then the interpretations of the EEG signal patterns have been a challenging topic. There are certain unwanted signals that are generated during the EEG signal recorded over the time that are called artifacts. These artifacts are later preprocessed as they produce interference. The artifacts are basically separated into two types: physiological [9] and nonphysiological [10]; the former occur due to the motion of the patient's body like eye movement, respiration, blinking of the optics, and the later take place due to some trouble in the EEG machine like electrode not placed properly, environment sources. Different types of feature reduction techniques are used to remove these artifacts. The wave forms generated in EEG signals are of various characters as identified infra.

1.3.1 Delta Wave (δ)

It has a frequency of 3 Hz or lower. By nature it is the highest in amplitude and the slowest waves. It is normally seen in infants up to 1 year. It occurs focally with subcortical lesions and in general distribution with diffuse lesions, metabolic encephalopathy hydrocephalus, or deep midline lesions.

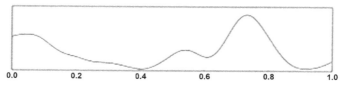

Fig. 1.4 Delta signal wave.

It is generally most prominent frontally in case of adults (e.g., FIRDA—Frontal intermittent rhythmic delta) and posterior in case of children (e.g., OIRDA—Occipital intermittent rhythmic delta). Fig. 1.4 shows a sample delta wave.

1.3.2 Theta Waves (θ)

It is having a frequency range between 3.5 and 7.5 Hz and is generally considered as *slow* activity by nature. It is perfectly normal in children up to 13 years and in sleep, but abnormal in awake adults. It can be viewed as a demonstration of focal subcortical lesions; it can as well be seen in generalized distribution in diffuse disorders such as metabolic encephalopathy or some cases of hydrocephalus. Fig. 1.5 shows a sample theta wave.

1.3.3 Alpha Waves (α)

It is having a frequency between 7.5 and 13 Hz. It generally occurs in the posterior portion of the head on both the sides. It is having higher amplitude on the dominant side. It generally occurs when closing the eyes and relaxing, and disappears when eyes are open or person in alert state by thinking or calculating. It is the major form of EEG signal generally seen in normal relaxed adults. Fig. 1.6 shows a sample alpha wave.

1.3.4 Beta Waves (β)

Beta waves are very fast in action. It has a frequency range of 14 Hz and greater. It is most frequently picked up along two sides in symmetrical distribution and is most evident frontally. It generally occurs due to sedative-

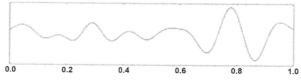

Fig. 1.5 Theta signal wave.

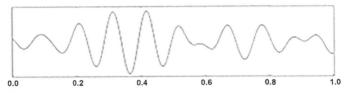

Fig. 1.6 Alpha signal wave.

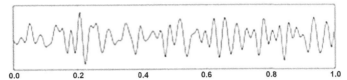

Fig. 1.7 Beta signal wave.

hypnotic drugs, especially the benzodiazepines and the barbiturates. It may be missing in fields of cortical damage. It is mostly seen as a normal brain wave pattern. It is the dominant brain wave in case of patients who are alert or anxious or have their eyes open. Fig. 1.7 shows a sample beta wave.

1.3.5 Gamma Waves (γ)

It has the frequency range approximately 30–100 Hz. Gamma waves generally represent combination of different nerve cells together into a network for the intent of stocking out a certain cognitive or motor part. Fig. 1.8 shows a sample gamma wave.

1.3.6 Mu Waves (μ)

It has a range between 8 and 13 Hz and partially overlaps with other type of brain waves. It reflects the synchronous firing of motor neurons in rest state. Fig. 1.9 shows a sample mu wave.

The International Federation of Societies for Electroencephalography and Clinical Neurophysiology has recommended the 10–20 system of

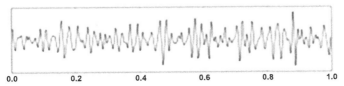

Fig. 1.8 Gamma signal wave.

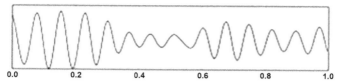

Fig. 1.9 Mu signal wave.

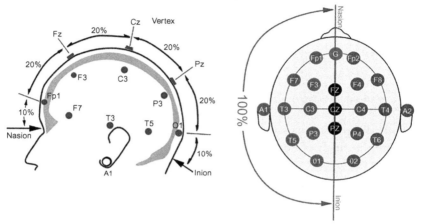

Fig. 1.10 The 10/20 system—standardized placement of electrodes on scalp for EEG measurements.

electrode placement for clinical EEG recording, which is schematically illustrated in Fig. 1.10.

The name 10–20 indicates the fact that the electrodes along the midline are placed at 10%, 20%, 20%, 20%, 20%, and 10% of the total nasion–inion distance; the other series of electrodes are also placed at similar fractional distances of the corresponding reference distances [11]. The interelectrode distances are equal along any anterior-posterior or transverse line, and electrode positioning is symmetrical. Each side has a letter to identify the lobe and a number to identify the hemisphere location. The letters F, T, C, P, and O stand for frontal, temporal, central, parietal, and occipital lobes, respectively. A z (zero) refers to an electrode placed on the midline. Even numbers (2, 4, 6, and 8) refer to electrode positions on the right hemisphere, whereas odd numbers (1, 3, 5, and 7) refer to those on the left hemisphere. The letter codes A, Pg, and Fp identify the earlobes, nasopharyngeal, and frontal polar sites, respectively. The typical EEG instrumentation is based along the low-pass filter with cutoff frequency of 75 Hz. Automated analysis of the EEG is beneficial for neurodiagnosis and the evaluation of treatment alternatives, if available, in the area of medical and neuroscience. All the same, automated EEG analysis is nonetheless a really challenging problem due to the complexity of extracting useful information from EEG signals. EEG signal processing requires dealing with multidimensional data. As a consequence, there is a need to minimize the dimension of the problem (features/channels) to ameliorate the operation. ML deals with programs and procedures that determine from experience (supervised learning), that is, programs that improve or adjust their operation at every step by iterative manner on a certain project or group of tasks over time.

1.4 OVERVIEW OF ML TECHNIQUES

ML is a type of artificial intelligence (AI) technique which provides computers with the ability to learn without being explicitly programmed. ML focuses on the evolution of computer programs that can instruct themselves to adjust to the environment so that they can get exposed to fresh information. It learns automatically how to arrive at accurate predictions based on past observations. The goal of ML is to design programs that read and/or disclose, that is, automatically improves their operation on certain chores and/or conform to changing circumstances over time. The effect can be a learned program which can take out the chore it was designed for or a learning program that will always improve and adjust. It classifies examples into giving set of categories as identified in Fig. 1.11.

When applied to a real-world application, ML involves the following basic steps as shown in Algorithm 1.1.

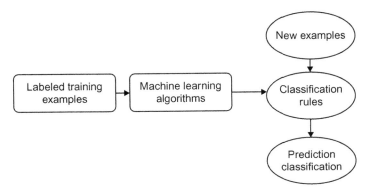

Fig. 1.11 Procedure for ML algorithms.

Algorithm 1.1 Machine Learning

Step 1: *Identifying the problem.*

Step 2: *Collecting and identifying the dataset.*

Step 3: *Preprocessing of data.*

Step 4: *Defining the training dataset.*

Step 5: *Selecting a model or a classifier.*

Step 6: *Selecting the training algorithm for the problem.*

Step 7: *Repeat from step 8 and step 9.*

Step 8: *Train the dataset to examine fitness using the training algorithm for the classifier.*

Step 9: *If found satisfactory or if the output class labels found to be correct for the input class label, then got to step 10 else tune the algorithm parameters and go to step 7.*

Step 10: *Evaluate the prepared model with the test dataset.*

ML algorithms are essential tools for an online classification of linear and nonlearning tasks [12]. Different methods are there for processing the EEG signals such as frequency domain, time domain, and time-frequency domain. Here, ML techniques are used for classification of EEG signal for epileptic seizure identification. ML is the application of algorithms to extract patterns from EEG signals [13]. However, there are other steps that are carried out, for example, data cleaning or preprocessing, data reduction or projection, incorporation of prior knowledge, proper validation and interpretation of results while analyzing EEG signals. EEG analysis has number of challenges which makes it suitable for ML techniques [14] like

1. EEG signals have large temporal variance.
2. EEG comes in large databases.
3. EEG recordings are very noisy.

EEG data can be analyzed using classification, clustering, regression, sequence analysis, and so on. The most common application for EEG today is in the field of enhancing the clinical research areas related to BCI for monitoring the various states of brain detecting the neural states and disorders. Applications of ML in BCI have shown a tremendous change leading to improvement in the field of clinical research and medical science.

The different ML techniques under the umbrella of ANNs used in this research work are multilayer perceptron neural network (MLPNN), radial basis function neural network (RBFNN), probabilistic neural network (PNN), and recurrent neural network (RNN). Other popular ML technique widely used in the task of classification is SVM.

1.4.1 Multilayer Perceptron Neural Network

ANN simulates the operation of a neural network of the human brain and solves a problem. Generally, single-layer perceptron neural networks are sufficient for solving linear problems, but nowadays the most commonly employed technique for solving nonlinear problems is MLPNN [15]. It can hold various layers such as one input and one output layer along with one or more number of hidden layers. There are connections between different layers for data transmission. The connections are generally weighted edges to add some extra information to the data where it can be propagated through different activation functions. The heart of designing MLPNN is the training of network for learning the behavior of input–output patterns. In this work, we have designed a MLPNN using Java. This network is trained with the help of three popular training algorithms such as back-propagation (BP) [16], resilient-propagation [17], and Manhattan update rule and quick propagation.

1.4.2 Radial Basis Function Neural Network

RBFNN is a type of feed-forward network trained using a supervised training algorithm. The main advantage of RBFNN is that it has only one hidden layer. The RBF network usually trains faster than BP networks. This kind of network is less susceptible to problems with nonstationary inputs because of the behavior of radial basis function hidden units.

1.4.3 Recurrent Neural Network

RNN [18] is a special type of ANN having a fundamental feature, that is, the network contains at least one feedback connection [19], so that activation can flow round in a loop. This feature enables the network to do temporal processing and learn the patterns. The most important common features shared by all types of RNN [20,21] are they incorporate some form of multilayer perceptron as subsystem and they implement the nonlinear capability of MLPNN [22,23] with some form of memory.

1.4.4 Probabilistic Neural Network

PNN was first proposed by Specht in 1990. It is a classifier that maps input patterns in a number of class levels. It can be forced into a more general function approximation. This network is organized into a multilayer feed-forward network with input layer, pattern layer, summation layer, and the output layer. PNN [22] is an implementation of a statistical algorithm called *kernel discriminant analysis*. The advantages of PNN are as follows: it has a faster training process as compared to BP. Also, there are no local minima issues. It has a guaranteed coverage to an optimal classifier as the size of the training set increases. But it has few disadvantages like, slow execution of the network because of several layers and heavy memory requirements, and so on.

1.4.5 Support Vector Machines

SVM is the most widely used ML technique-based pattern classification technique available nowadays. It is based on statistical learning theory and was developed by Vapnik in the year 1995. The primary aim of this technique is to project nonlinear separable samples onto another higher dimensional space by using different types of kernel functions. In late years, kernel methods have received major attention, especially due to the increased popularity of SVM [24]. Kernel functions play a significant role in SVM [25] to bridge from linearity to nonlinearity. Least square SVM [26] is also an important SVM technique that can be applied for classification task [27]. Extreme learning machine and Fuzzy SVM [28–30] and genetic algorithm tuned expert model [28] can also be applied for the purpose of classification. In this analytical work, three different types of kernel functions [31], that is, linear, polynomial, and RBF kernel [32] were evaluated.

1.5 SWARM INTELLIGENCE

Inspired by studies in neurosciences, cognitive psychology, social ethology, and behavioral sciences, the concept of swarm intelligence (SI) [33] was introduced in the domain of computing and AI in 1989 [34] as an innovative, collective, and distributed intelligent paradigm for solving problems, mostly in the domain of optimization, without centralized control or the provision of a global model. SI is based on the collective behavior of decentralized, self-organized systems. It may be natural or artificial. Natural examples of SI are ant colonies, fish schooling, bird flocking, bee swarming, and so on. Besides multirobot systems, some computer program for tackling optimization and data analysis problems are examples for some human artifacts of SI.

The most successful SI techniques are particle swarm optimization (PSO) and artificial bee colony algorithm (ABC). In PSO, each particle flies through the multidimensional space and adjusts its position in every step with its own experience and that of peers toward an optimum solution by the entire swarm. Therefore the PSO algorithm is a member of SI. PSO method was first introduced in 1995 [35,36]. Since then, it has been used as a robust method to solve optimization problems in a wide variety of applications. On the other hand, the PSO method does not always work well and still have room for improvement. This book discusses a conceptual overview of the PSO algorithm and an improvement over the basic PSO. Besides this, it also describes a modified ABC algorithm to enhance the performance of RBFNN.

The ABC algorithm was first proposed by Karaboga in 2005, by simulating the foragers' behavior of finding food sources. Three types of bees are adopted in this algorithm, and which itself have different division of works. Further, these techniques have been described in detail in Chapters 4 and 5 along with the modifications performed on the existing algorithm.

1.6 TOOLS FOR FEATURE EXTRACTION

In this work, the data was collected from Bonn University site [37] which is a publicly available database used by Andrzejak et al. [38] related to diagnosis of epilepsy. This resource provides five sets of EEG signals. Each set contains reading of 100 single channel EEG segments of 23.6s duration each. These five sets are described as follows. Datasets A and B are considered from five healthy subjects using a standardized electrode placement system. Set A contains signals from subjects in a slowed down state with eyes open. Set B also contains signal same as A but with the eyes closed. The datasets C, D, and E are recorded from epileptic subjects through intracranial electrodes for interictal and ictal epileptic activities. Set D contains segments recorded from within the epileptogenic zone during seizure-free interval. Set C also contains segments recorded during a seizure-free interval from the hippocampal formation of the opposite hemisphere of the brain. Set E only contains segments that are recorded during seizure activity. All signals are recorded through the 128-channel amplifier system. Each set contains 100 single-channel EEG data. In all, there are 500 different single-channel EEG data. All these data collected are in signal form, that is, numerical values in a single-column format. After data collection, different techniques have been reviewed for feature extraction. It has been found that DWT is the most efficient and powerful technique for analyzing EEG signal that is in one-dimensional format. Hence, MATLAB is being used for implementing this technique and the results have been obtained. After feature extraction different ML techniques have been applied to study the behavior of these techniques over the classification task on EEG signal in epilepsy identification. All these techniques have been implemented using Java platform (JDK 1.8) and Eclipse Mars IDE.

   Basically, all types of signal generated under medical diagnosis are analyzed in time domain with their amplitudes. Like ECG, EEG signals are generally collection of amplitudes with respect to time. The pathological condition of a patient can be observed from the graph plotted using the data. If there is any significant deviation in shape it can be shown and observed properly by visualizing the graph [39]. But sometimes it is necessary to get the frequency content of a signal for proper and accurate analysis of a signal. It can be done by using any transformation technique such as Fourier Transform. But the disadvantage is that it is not so effective for transient signals such as EEG. Because EEG signal has very uncertain and rapidly changing frequency, it is very difficult to analyze the signals effectively. Hence, some other transformation technique is needed such as wavelet transformation to analyze EEG signals [39]. This is just a new perspective for analysis and processing of signals. The basic idea behind this technique is to use a scale for analysis. The wavelet transform can be divided into two categories like continuous wavelet transform (CWT) and DWT [40]. CWT was first developed as an alternative to short time Fourier transform. In this technique, the signal is transformed by calculating the product of the signal with a function which is called as wavelet function [41]. This transform is again calculated for different time domain. The wavelet function is defined in Eq. (1.1).

$$\text{CWT}\,(a, b) = \int\limits_{-\infty}^{\infty} x(t) * \varphi_{a,b}^{\nabla}(t)\,dt \qquad (1.1)$$

where, $x(t)$ represents the original signal. a, b represent the scaling factor and translation along the time axis, respectively. The symbol ∇ denotes the complex conjugation and $\varphi_{a,\,b}^{\nabla}$ is calculated by scaling the wavelet at time b and scale a.

$$\varphi_{a,b}(t) = \frac{1}{\sqrt{a}} \varphi\left(\frac{t-b}{a}\right) \qquad (1.2)$$

   where $\varphi_{a,\,b}(t)$ represents the mother wavelet. In CWT, it is assumed that the scaling and translation parameter a and b change continuously. But the main disadvantage of CWT is the calculation of wavelet coefficients [18] for every possible scale can result in a large amount of data. It can be overcome by the help of DWT. It analyses the signal at different frequency bands by decomposing the signal into a set of high- and low-pass filters called as *approximation* and *detailed coefficients*. These coefficients can be calculated by using the wavelet toolbox available in MATLAB. Using the predefined functions available inside this toolbox we can easily extract the features of EEG signal. Figs. 1.12–1.14 describe the EEG signal decomposition using DWT with Daubechies wavelet of order 2 up to level 4.

From the data available at Ref. [37], a rectangular window of length 256 discrete data was selected to form a single EEG segment. The wavelet coefficients have been computed using Daubechies of order four. This technique was found to be more suitable because of its smoothing features which are more appropriate to detect changes in EEG signal.

In our work, the original signal has been decomposed as four detailed coefficients (d_1, d_2, d_3, d_4) and four approximation coefficients (a_1, a_2, a_3, a_4). For simplicity, all the approximation coefficients are ignored except the one in the last step, that is, a_4. Hence, the signal is decomposed into five segments by using DWT. In this work, for four detailed coefficients we get 247 coefficients ($129 + 66 + 34 + 18$) and 18 for approximation coefficients. Several statistical features have been extracted. But for this study, four important features were taken into considerations:

a. Maximum of wavelet coefficients in each subband.
b. Minimum of wavelet coefficients in each subband.
c. Mean of wavelet coefficients in each subband.
d. Standard deviation of wavelet coefficients in each subband.

Therefore for five coefficients all total 20 features have been extracted and the dataset have been constructed. Like DWT, there are many other techniques for extracting features from an EEG dataset. The features that can be extracted from the techniques that include

a. Fractal dimension (Higuchi and Petrosian)
b. Hurst exponent
c. Spectral, approximation, and SVD entropy
d. Detrended fluctuation analysis
e. Hjorth mobility and complexity

These extracted features provide us a sector to explore the EEG dataset in a more detailed way for the purpose of classification. Hence, along with the features that have been extracted from DWT, we have also used these nine features for our experimental analysis process of classification. The techniques have been elaborated and stated as following. Let us consider the signal for which features are extracted is represented as follows:

$$X = [x_1, x_2, x_3, \ldots, x_N]$$

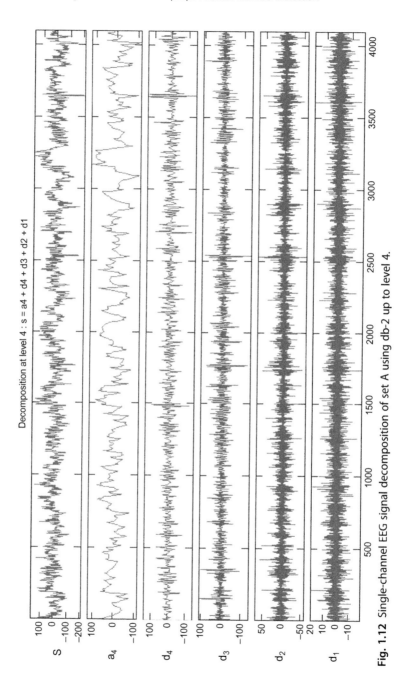

Fig. 1.12 Single-channel EEG signal decomposition of set A using db-2 up to level 4.

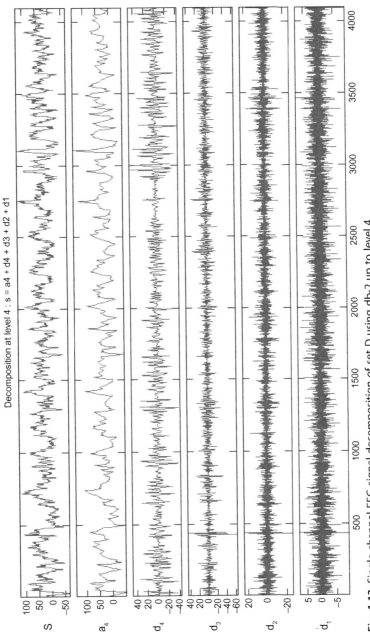

Decomposition at level 4 : s = a4 + d4 + d3 + d2 + d1

Fig. 1.13 Single-channel EEG signal decomposition of set D using db-2 up to level 4.

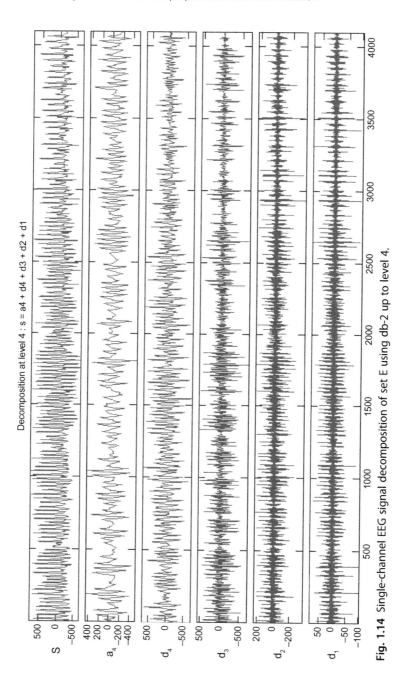

Fig. 1.14 Single-channel EEG signal decomposition of set E using db-2 up to level 4.

a. Fractal dimension [42, 43]

It is one of the important features of a signal that may contain some information about the geometrical shape at different scales. These information can be extracted using different methods such as proposed by Petrosian and Higuchi and named accordingly Petrosian fractal dimension (PFD) and Higuchi fractal dimension (HFD). Eq. (1.3) shows formula for calculating PFD.

$$PFD = \frac{\log_{10} S}{\log_{10} S + \log_{10}\left(\dfrac{S}{S + 0.4 * S_\varphi}\right)} \tag{1.3}$$

where S is the series length and S_φ is the number of sign changes in the signal.

Similarly, HFD is the slope of line that best fits the curve of $\ln(Z(k))$ and $\ln(1/k)$. Where $Z(k)$ is defined in Eq. (1.4).

$$Z(k) = \frac{\sum_{i=1}^{k} Z(i, k)}{k} \tag{1.4}$$

$$\text{where } Z(m, k) = \frac{\sum_{i=2}^{(N-m)/k} |x_{m+ik} - x_{m+(i-1)k}| (N-1)}{\lfloor (N-m)/k \rfloor k}$$

This algorithm constructs k new series from original series as follows:

$$x_m, x_{m+k}, x_{m+2k}, \ldots, x_{m+\lfloor (N-m)/k \rfloor k}, \text{ where, } m = 1, 2, \ldots, k$$

b. *Hurst Exponent* [44]

It is generally used as a measure of long-term memory of time series data. It can be calculated by first calculating deviation from mean of time series and then by calculating the rescaled range statistics (R/S). First, we need to calculate the accumulated deviation from mean of time series within range T as shown in Eq. (1.5).

$$X(t, P) = \sum_{i=1}^{t} x_i - \overline{x}, \text{ where,} \overline{x} = \frac{1}{P}\sum_{i=1}^{P} x_i, \quad t \epsilon [1 \ldots N] \tag{1.5}$$

Then $R(P)/S(P)$ is calculated as per the formula shown in Eq. (1.6).

$$\frac{R(P)}{S(P)} = \frac{\max\left(X(t, P)\right) - \min\left(X(t, P)\right)}{\sqrt{1/P \sum_{i=1}^{P} [x_t - \overline{x}]^2}} \tag{1.6}$$

The Hurst Exponent is calculated as the slope of line produced by $\ln\left(\frac{R(P)}{S(P)}\right)$ versus $\ln(P)$.

c. Spectral, approximation, and SVD entropy [45–47]

Entropy can be measured as the spread of data. Data with broad or flat probability distribution has a high entropy and vice versa. This is one of the statistical descriptor of variations in EEG signal. Spectral entropy can be defined in terms of power spectral intensity and relative intensity ratio (RIR) as shown in Eq. (1.7).

$$SI = \frac{-1}{\log(K)} \sum_{i=1}^{K} RIR_i \log RIR_i \qquad (1.7)$$

$$\text{where, } RIR_j = \frac{PSI_j}{\sum_{k=1}^{K-1} PSI_k} \quad \text{and} \quad PSI_k = \sum_{i=\lfloor N(f_k/f_s)\rfloor}^{\lfloor N(f_k+1/f_s)\rfloor} |X_i|,$$

$$k = 1, 2, \ldots, K-1,$$

where f_s is the sampling rate, X_i denotes FFT of time series x_i, and f_1 to f_K represents K slices of frequency band of equal or unequal widths. Similarly, *approximation* entropy is a statistical parameter computed for a time series. SVD entropy defines an entropy measure by the help of singular value decomposition.

d. Detrended fluctuation analysis [48]

It is another important feature extracted for analysis of signals with scale invariant structure. It is a method for determining statistical self-affinity of a signal. The exponents obtained are almost similar to Hurst exponent.

e. Hjorth mobility and complexity [49]

Hjorth parameters generally describe the statistical properties of a signal. This is a very popular signal analysis method proposed by Hjorth in 1970, used for analyzing electroencephalogram signals. It has mainly three kinds of parameters such as activity, mobility, and complexity. In this domain, we have used the last two for analysis of EEG signal, which uses the activity parameter. Mathematically, it can be defined as shown in Eq. (1.8).

$$\text{Mobility} = \sqrt{B2/AVG} \quad \text{and} \quad \text{Complexity} = \sqrt{(B4*AVG)/(B2*B2)} \qquad (1.8)$$

$$\text{where AVG} = \sum x_i/N, \quad B2 = \sum d_i/N,$$

$$B4 = \sum \left((d_i - d_{i-1})^2/N\right), \quad d_i = x_i - x_{i-1}$$

After the features have been extracted, the next important task is to design a classifier model for classifying the seizure and nonseizure signals. For this task we have considered the RBFNN for its architectural simplicity and less number of parameters required for adjustment.

1.7 CONTRIBUTIONS

EEG analysis and classification is one of the most important tasks in automatic detection of any neurological diseases. There are lots of research works going on in this topic to address different issues. Our contribution for this book work has been categorized into following points.

o EEG signal analysis and feature extraction using DWT and other techniques.

Here, we are trying to discover or extract the hidden features present in an EEG signal. For this, we have used MATLAB. Through this programming tool we have extracted different statistical features such as:

o Maximum
o Mode
o Minimum
o Standard deviation
o Mean
o Mean absolute deviation
o Median
o Median absolute deviation
o Range
o L2 norm
o L1 norm
o Max norm

Few of them have been considered for our research work based on our extensive literature survey. Other than DWT, some mathematical feature extraction techniques such as fractal dimension; spectral, approximation, and SVD entropy; Hurst exponent; detrended fluctuation analysis; Hjorth mobility and complexity, and so on, have been used. For all this analysis, a publicly available preprocessed dataset for epileptic seizure identification have been used.

o After all the features have been extracted, an empirical analysis is performed for different ML techniques for classification of EEG signal to detect epileptic seizure

The performance of different ML techniques like ANNs and its variants such as RBFNN, RNN, and PNN in addition to SVM was investigated to classify EEG signal for epilepsy identification.

o   The next task is to enhance the performance of RBFNN. To improve the accuracy, we have introduced an effective hybridization of RBFNN with a modified version of PSO technique

We have proposed a very novel technique of training and optimizing parameters of RBFNN by integrating a modified version of PSO algorithm. The modification in PSO is specifically done for increasing search speed around the global optimum. We have introduced an innovative technique to change the value of inertia weight linearly and nonlinearly rather than taking a constant value always.

At last we have further enhanced the performance of previous technique by replacing PSO algorithm with a modified version of ABC algorithm. The modification in ABC has been done in the initialization and selection stage. The experimental evaluation shows that there is a remarkable increase in the performance of RBFNN for classification of EEG signal in epileptic seizure identification.

1.8 SUMMARY AND STRUCTURE OF BOOK

In this research work, we have started our journey from an extensive literature review on the area of EEG signal analysis and classification for epileptic seizure identification to bridge the research gap by developing a few novel models.

In a nutshell the basic structure and organization of human brain and a generation of EEG signal has been studied. Also study has been performed along the recording methodologies of EEG signal in human brain. This survey has revealed the behavior of EEG and thus gave insights into the process of analysis and classification tasks.

The rest of the chapters of this book is set out as follows. Chapter 2 describes the extensive literature survey done on EEG signal analysis, preprocessing, and classification using ML techniques. Chapter 3 provides an empirical study on the performance of different ML techniques in EEG signal classification for epilepsy identification. This chapter also describes about different experimental studies performed and different parameters involved. Chapter 4 describes about our proposed novel technique for accuracy improvement of RBFNN in classification of EEG signal using SI-based optimization techniques such as PSO. Chapter 5 describes about our proposed technique for accuracy improvement of RBFNN in classification of EEG signal using a modified version of ABC algorithms and comparison with previous technique. Finally in Chapter 5, we conclude with some constraints along with few insights about the future work.

CHAPTER 2

Literature Survey

This chapter introduces the literature surveys done in the field of EEG signal analysis and feature extraction, along with the different ML techniques applied in the classification of EEG signal. Our research work primarily focuses on EEG signal classification for epileptic seizure identification. This work is generally performed to distinguish between epileptic seizures from nonepileptic seizures (NES). It is very important from the medical point of perspective because the two types of seizures are completely dissimilar from each other although the behavior of both is same. Treatment for both is also entirely unlike. Epileptic seizure generally originates from our brain, but nonepileptic ones don't originate from the mind. They are generated due to any other medical problems in our body such as high blood pressure, high carbohydrate, and so on. Hence, it is an important problem to be accosted by the computer researchers. Because an automated strategy for this categorization is required, many researchers have been getting at this focal point to explore some more effective ways to analyze and classify EEG signal for epilepsy identification. Other than epilepsy, there are other neurological diseases such as coma; brain death has also been addressed by the researchers. Some other most important application of EEG analysis is in the case of BCI. This is a very sound and a powerful communication channel between human brain and computers. This system basically helps physically disabled individuals to communicate with computing machines. Through this interface a person can operate a computer without physically moving his or her body parts. This interface is designed so intelligently that it can read the EEG brain signals and accordingly operate the machines. Several intensive researches are going on in this area to develop this interface. The most important prospect of all these researches is based on the recording of EEG signal. There are different ways developed by the scientists to record these signals. One of the standardized systems for EEG recording is 10–20 international standard system. This is an invasive medical procedure where no medical surgery is needed. It simply places the electrodes on the human scalp according to certain rules defined in the standard. This electrode placement is also called as *extra cranial electrode placement*. Another process of electrode placement is called as intracranial *electrode placement*. In this process, electrodes are placed within the human brain through some surgical process to record the EEG thoroughly. In this study, we have confined ourselves to the EEG signal analysis for epileptic seizure identification. Hence, our survey is restricted to EEG signal analysis along with the preprocessing task in general to epileptic seizure identification in specific.

2.1 EEG SIGNAL ANALYSIS METHODS

EEG signal analysis for feature extraction is one of the most significant tasks in epilepsy identification. In this process, we have to conduct a well-specified and well-structured signal analysis technique to press out the statistical features from them. This procedure is necessary for the requirement of classification tasks. There are various signal analysis techniques that are already available. Some of them are fast Fourier transform techniques (FFT), STFT, CWT, DWT, and so on. Fourier transform of a signal in time domain is taken for frequency amplitude representation of that signal. We can also take the signal in frequency domain because some information that is not visible in time domain may be seen in the frequency domain. Although FFT is the most widely used transform technique, it is not the only one available. Other transformation techniques may include STFT, Hilbert transform, Wigner distributions, Radon transform, and wavelet transform. One of the disadvantages of FFT is that it cannot analyze the signal in time and frequency domain simultaneously. That implies it cannot yield the information as at what time these frequency components exist. This transformation is not required when the signal is stationary. Stationary signals are those signals whose frequency does not change according to time. Therefore for stationary signals it is not required to know at what time a particular frequency component exists. For stationary signals the frequency component which exists in the signal continues to remain for a total duration of that signal. In this transformation technique we generally pass a whole signal into a high- and a low-pass filter. These filters generally break the signals into groups of high- and low-frequency signals and this process continues for a certain number of times. This process is called as decomposition. FFT decomposes a signal into different complex exponential functions as presented in Eqs. (2.1), (2.2).

$$S(f) = \int_{-\infty}^{\infty} s(t) * e^{-2j\pi ft} dt \qquad (2.1)$$

$$s(t) = \int_{-\infty}^{\infty} S(f) * e^{2j\pi ft} df \qquad (2.2)$$

where t stands for time, f for frequency, s denotes signal in time domain, and S denotes signal in frequency domain. As we have stated earlier, FFT is only suitable for stationary signals whose frequency component does not change according to time. For this reason this technique is not desirable for nonstationary signals. However, we have discussed in Chapter 1 that EEG signals are very transient or nonstationary in nature, so FFT is not suitable for this analysis. To surmount this problem researchers have projected a variety in FFT that is called as STFT. In STFT a small portion of nonstationary signal is considered as stationary. In this technique, we have to take a window whose width depends on the property of nonstationary of the signal. If the signals are more transient, then we have to take a narrow window to make that portion stationary. For this we have to select a window function. The width of this window must be equal to the different segments of that signal where the stationary property of that signal can be obtained. The specific equation for STFT is given in Eq. (2.3).

$$S(t,f) = \int_{t} [s(t) * \omega(t - t')] * e^{-2j\pi ft} dt \qquad (2.3)$$

where $\omega(t)$ is the window function used for STFT. So, from Eq. (2.3) it is clear that STFT is nothing but same as FFT multiplied with window function. The window functions are generally taken as Gaussian function. One of the major problems with STFT is the *Heisenberg's Uncertainty Principle*. One cannot say exactly what frequency content exists at what time component. For this problem another type of signal analysis technique has been introduced called as multiresolution analysis. It analyzes the signals with different frequencies at different resolutions. Hence, the resolution is not same for all type of frequencies. For high frequencies a good time resolution and poor frequency resolution have been provided. Similarly, for low frequencies a good frequency resolution and poor time resolution have been provided. Also, to overcome the shortcomings of STFT, the Wavelet Transform technique has been introduced. The term wavelet means a small wave whose width depends on the parameters of a mother wavelet. Wavelet transform is of two types:

o   CWT
o   DWT

The working procedure of STFT and WT approach is same. That is, a function multiplied with the signal called as *window function* in STFT and *mother wavelet* in WT. However, the difference between them is in CWT, the width of the window is changed during transformation. The CWT can be defined as follows.

$$\text{CWT}(a, b) = \frac{1}{\sqrt{|b|}} \int s(t) * \varphi\left(\frac{t - a}{b}\right) dt \tag{2.4}$$

here *a*, *b* are called as translation and scale parameters. The φ function is called as the *mother wavelet*. Again, the most important question raised on CWT is that can it be able to transform the signal in discretized time and scale parameters. The answer is no for which we have to go for a new technique called as DWT. Although the discretized version of CWT is possible for discrete data analysis, it is not truly discrete. DWT provides the most efficient way to analyze the signal as well as reduction in computational time. DWT is nothing but the decomposition of a discrete time signal in time-frequency domain. The difference between DWT and CWT is along with the different scales, different cutoff frequencies are used for analysis of signal in the former technique. The signal is passed through a group of low- and high-pass filters to analyze low and high frequencies, respectively. Filtering is nothing but a mathematical operation of convolution. Let *lp(n)* represents impulse response of a low-pass filter, and *s(n)* represents the signal itself. Then the convolution operation can be represented as given in Eq. (2.5).

$$s[n] * lp[n] = \sum_{i=-\infty}^{\infty} s[i] * lp[n - i] \tag{2.5}$$

Similarly, a high-pass filter is used to analyze the high frequency content. The scale is changed either by up-sampling or down-sampling. This technique is also called as subsampling operation. Fig. 2.1 describes the decomposition of a signal using DWT with down-sampling up to certain number

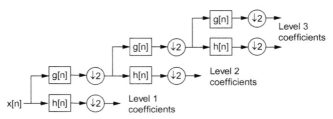

Fig. 2.1 Signal decomposition using DWT.

of levels. Each level gives rise to two different coefficients called as detailed and approximation coefficient.

The main difference of DWT from FFT is that the time localization of different frequency components is preserved. However, this localization depends on at which level they exist. It may occur in high frequency or in low frequency. One of the most important properties of DWT is that the bands of frequencies are not totally independent of each other. The relationship can be established using Eq. (2.6).

$$hp[P-1-n] = -1^n * lp[n] \qquad (2.6)$$

where $hp[n]$ is the high-pass filter, $lp[n]$ is the low-pass filter, and P is the filter length. Another important aspect of DWT is the selection of appropriate mother wavelet function. There are various choices available. Some of them are Daubechies wavelet, Haar wavelet, Coiflet wavelet, Morlet wavelet, Mexican-Hat wavelet, and so on.

In this part of literature survey, we have studied different methods adopted by different researchers for the analysis and feature extraction from EEG signal. Gandhi et al. [50] have analyzed the EEG signal for classification purpose. They decomposed the signal using DWT with different wavelet functions such as Haar, Daubechies of order 2 to 10, Coiflet of order 1 to 10, and Biorthogonal of orders 1.1, 2.4, 3.5, and 4.4. The different features have been extracted such as energy, entropy, and standard deviation at different subbands. After all analysis task, they performed the classification task using PNN and compared with SVM. They found the features extracted from Coiflet of order 1 are more accurate for the classification task. Li et al. [51] have proposed a very effective multiwavelet-based time-varying scheme for analysis in time-frequency domain for EEG signals. After that orthogonal least square technique has been applied for sparse model selection and estimation of parameters. In their work, they have approximated time–dependent parameters using basis functions of *Multiresolution B-spline* wavelet. Guo et al. [52] have performed an automatic feature extraction using genetic programming. The main aim of this work is to improve the performance of classifier and reduce the dimension of features simultaneously. There are many research works that have been performed in this field of feature extraction using genetic programming. Raymer et al. [53] also applied the genetic programming to improve the performance of KNN classifier without feature reduction. Sherrah [54] also created an evolutionary computation-based feature extraction methodology. Tzallas et al. [55] also developed a genetic programming-based tree to process the features for classification. Acharya et al. [56] have proposed an automatic method for epilepsy identification from EEG signal by extracting different entropy features from EEG signal. These features include approximation entropy, sample entropy, phase entropy 1, and phase entropy 2. Approximation entropy is a measure for data regularity for a signal. High data regularity means less approximation entropy and vice versa. Sample entropy is almost similar to approximation entropy except sample entropy is independent of the record length. The two phase entropies that is 1 and 2 are same as spectral entropy. Nasehi et al. [57] have proposed a dimensionality reduction technique for features using general tensor discriminant analysis. In their work, the EEG signal was decomposed as spectral, spatial, and temporal domain using Gabor function of order three tensors. Oveisi [58] has used independent component analysis (ICA) as a signal processing technique. It is used to distinguish between different mental tasks in BCI system. Also, this is evaluated through some classification task. Delgado Saa and Gutierrez [59] have proposed a method for feature extraction and classification of EEG signal in BCI system. First, they extracted the features using parametrical methods for the calculation of power spectral density and Fisher criterion for separability. Then *linear discriminant analysis* was used for the classification of the signal. Subha et al. [60] have performed an extensive survey on EEG signal analysis. They studied almost every process available for analysis of EEG signals for different applications. According to Pradhan et al. [61], linear prediction can be used for storage and transmission of EEG signal. Tzyy-Ping et al. [62] have used ICA for removal noises from EEG signal and separate individual sources of brain signals. Spectral analysis can be taken as an effective method of signal analysis. It can be categorized into two types such as parametric and nonparametric methods. Welch [63] has suggested the Welch method that is popularly used for estimation of power spectrum of a signal. This is one type of nonparametric method. One of the disadvantages of these methods is the leakage effect due to windowing. To avoid this problem we can go for parametric methods for spectral analysis. Parametric methods are also able to provide better frequency resolution. Among all the parametric methods Burg method is widely used. In this method the reflection coefficients are calculated without calculating autocorrelation. Among other signal analysis techniques STFT and Wavelet transform can be included.

2.2 PREPROCESSING OF EEG SIGNAL

Another important task which is performed during the signal analysis and classification is the preprocessing task. This task includes operations required for preconfiguring the signal for better performance in consequent stages. One of the popular tasks in EEG signal preprocessing is the artifacts removal. Generally, the origin of the EEG signals is the cerebrum. This is the main part of the brain where signals are generated. But there are some situations when the signals are generated from any noncerebral origin. These are called as artifacts. These are always assumed as unwanted signals or noise in the signal. These signals always contaminate the original signals. Hence for better analysis of the signals it is necessary to remove these noise or artifacts. Sometimes it is necessary for correct interpretation of EEG signals clinically. The artifacts are broadly classified as external and internal artifacts. External artifacts are caused by any external disturbances. Internal artifacts are generally caused due to the person itself. EEG recording is usually done by placing the electrodes on the human scalp connecting to some recording device. There may be some situations where there are any electrical disturbances in machineries. Due to this there may be disturbances caused in EEG recording. These are artifacts referred to as external artifacts. Internal artifacts are generally caused due to different body activities of a person. These activities may include patient's movement, eye blinks, muscle movements, and so on. Several researches have been performed in this area for efficient removal of noise from EEG signal for the development of brain computer interface system. Cheng et al. [64], Kubler et al. [65], and Fabiani et al. [66] have used techniques for noise removal called as common average referencing. surface Laplacian technique has been used by Babiloni et al. [67,68], Cincotti et al. [69,70], Qin et al. [71], Muller et al. [72], Millan et al. [73], and Schalk et al. [74]. Bayliss et al. [75], Erfanian et al. [76], Gao et al. [77], Peterson et al. [78], Wu et al. [79], Serby et al. [80], and Xu et al. [81] have used the ICA technique for preprocessing of EEG signal in BCI design. Chapin et al. [82], Guan et al. [83], Hu et al. [84], Isaac et al. [85], Lee and Choi [86], Yoon et al. [87], and Li et al. [88] have used principal component analysis (PCA) as a preprocessing technique for EEG signal in BCI design. Singular value decomposition is used by Trejo et al. [89] for preprocessing of EEG signal. Common spatio-spatial patterns is used as preprocessing for EEG by Lemm et al. [90]. Frequency normalization is used by Bashashati et al. [91], Borisoff et al. [92], Fatourechi et al. [93], and Yu et al. [94]. Peters et al. [95] used a technique called as local averaging technique for preprocessing of EEG. Robust Kalman filtering is a technique used by Makeig et al. [96]. Clark and Gonzalez [97] used common spatial subspace decomposition. It is also used by Li et al. [98], Liu et al. [99], and Wang et al. [100]. Wiener filtering is used by Vidal et al. [101]. Sparse component analysis is used by Zibulevsky et al. [102]. Pregenzer et al. [103] have used maximum noise fraction. Obeid and Wolf [104] used Spike detection methods. Neuron ranking methods have been used by Sanchez et al. [105].

2.3 TASKS OF EEG SIGNAL

After preprocessing of EEG signal several productive tasks can be performed on these preprocessed or refined data. The main aim of these tasks is basically to detect and solve any neurological disorder disease or help in developing human-computer interaction system. There are a lot of neurological diseases which are difficult to identify and cure. Hence, there must be an effective and advanced technique required for developing medical diagnostic systems in brain diseases. The greatest advantage of EEG signal recording is its speed. The recording speed is very high as compared to other recording techniques used for medical diagnosis. These signals can be recorded in a fraction of seconds or minutes. This recording generates a large volume of data within few seconds or minutes. EEG provides a very small spatial resolution as compared to MRI or PET. Hence, EEG is sometimes taken along with MRI for better medical diagnosis. There can be a lot of task performed by EEG analysis such as monitoring alertness, brain death, and coma state of a patient; locate different areas that may be damaged in the brain due to head injury or tumor; controlling the depth of anesthesia effects; identify epileptic seizure by distinguishing from NES; test the effect of drugs used for epilepsy patients; monitor human and animal brain developments; investigation of sleep disorder diseases; design of BCI, and so on. This research work basically focuses on one task of EEG that is epileptic seizure identification. But there are lots of research works going on in different aspects of EEG analysis. BCI is one of them, where a communication system has been designed for physically disabled persons to communicate with computers.

Zamir [106] has proposed a technique for epileptic seizure detection using linear least squares preprocessing. He has proposed a novel method for important hidden feature extraction to assist in epilepsy identification. Boashash and Ouelha [107] have performed a time-frequency-based feature extraction technique for analysis of a newborn EEG seizure detection. This analysis has been performed by the help of channel and feature fusion. He et al. [108] have used a Bayesian network with Gaussian distribution for the analysis of motor imagery EEG signal. They proposed this technique for multiclass classification problem. According to the authors in Ref. [107] this model performs better in comparison to other traditional models and reflects the distribution of EEG signal through creating a Gaussian model. Also, the work is not limited to their task of few numbers of channels because imagination activity is the combined result of all the regions in human brain. Ellenrieder et al. [109] have done a study on how synchronous activity of cortex is required for recording the EEG activities. Their investigation illustrates how sparse generator with random phase can add result with oscillatory scalp activity. Banerjee et al. [110] have done an investigation on brain dynamics by nonlinear analysis of music–induced EEG signal where the task was performed by using a unique method of EEG that is recorded by listening to music. Music is basically considered as a very important tool for analyzing human emotions. Hence, they considered the EEG recorded from human brain while listening to music in a relaxed state of mind. They also considered the readings from frontal lobe that is electrodes at F3 and F4. For feature extraction fractal dimension was used, that is, detrended fluctuation analysis. Also, different states of mind through EEG were considered that is during music and after music. Das and Bhuiyan [111] have proposed a classification technique to discriminate focal and non-focal EEG using entropy-based features extracted from empirical mode decomposition (EMD)–DWT technique. The features were extracted from different entropy-based techniques such as Shannon entropy, low-energy entropy, and Renyi entropy. Other than that, they used SVM as a classification technique and compared the performance with k–nearest neighbor classifier. Fu et al. [112] have proposed a technique for automatic detection of seizures through EEG analysis by considering Hilbert marginal spectrum analysis. Fu et al. also used EMD for feature extraction. They considered the different features like spectral entropy and energy features. Finally, SVM was used for classification task. Cuellar et al. [113] have performed a time-frequency observation of the EEG μ waves to monitor sensorimotor integration during swallowing that was discovered by bilateral clusters using ICA for sensorimotor μ components. Gao et al. [114] have performed an extensive study on the effect of different electrodes and electrode gels in case of MRI that was compared and the effect of three electrodes and their possible combinations on the signal-to-noise ratio of MRI was depicted. Peker [115] has performed a research to develop an effective sleep scoring system using ML techniques that he used for the automatic sleep scoring system to investigate sleep disorder disease. For this, they have used a complex valued nonlinear features and complex valued neural networks. From these features they have developed the automatic sleep scoring technique. These attributes are fed to a neural network to detect sleep disorders. Wang et al. [116] have performed an investigation of characteristics and functions of a human brain during Alzheimer's disease. They established a connection between different regions of brain with correlation to perform the above task. They used the techniques such as limited penetrable visibility graph, phase space method, and so on.

2.4 CLASSICAL VS MACHINE LEARNING METHODS FOR EEG CLASSIFICATION

This section discusses about the difference between classical methods and ML methods for classification task. Classical methods are generally considered as statistical modeling techniques. There is a very huge gap in between them according to their nature of solving problems. The formal definitions of these two are as follows. Statistical modeling is nothing but formulation of relationships among variables in terms of different mathematical equations. ML is a technique that can make a model able to learn from data without any explicit rule-based programming. Statistical method for solving classification problem is a subfield of mathematics that deals with calculating the relationship between variables to classify or predict an output. But ML is a subfield of computer science that deals with creating models that can learn from data without explicitly programming the model. Statistical modeling methods are very old and yet powerful techniques, whereas ML techniques are very new to problem-solving field. ML came into existence in the 1990s as a powerful and efficient problem-solving technique. Statistical modeling works on a number of premises. For example, a linear regression assumes a linear relation between independent and dependent variable, mean of error at zero for every dependent value, independence of observations, error should be normally distributed for each value of the dependent variable.

Similarly logistic regressions come with their own set of presumptions. Even a nonlinear model has to comply with a continuous segregation boundary. ML algorithms do take over a few of these things, but in general are spared from most of these premises. The greatest advantage of using a ML algorithm is that there might not be any continuity of boundary as established in the case study in a higher place. Likewise, we need not determine the distribution of dependent or independent variable in a ML algorithm. Fig. 2.2 shows the Venn diagram for coverage of ML and statistical modeling techniques in the area of classification.

The different naming conventions are also used for two different techniques. For example, models in statistical technique are called as network or graph in ML technique. We have performed an extensive literature survey

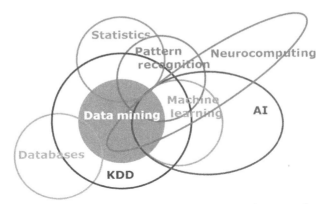

Fig. 2.2 Venn diagram for coverage of machine learning and statistical model in data mining.

on ML techniques used for EEG signal classification in the field of BCI design or epilepsy identification. Details of ML techniques in epilepsy classification have been discussed in Section 2.5.

Orhan et al. [117] have performed the EEG signal classification using two ML methods such as k-means clustering and MLPNN. Feature extraction has been performed using DWT. Then the wavelet coefficients are being clustered using k-means clustering algorithm for each frequency subband. Then MLPNN has been used for the classification task. Tangkraingkij et al. [118] have used ICA for preprocessing of EEG signal. Other than that, a supervised neural network model was used for identification of a person. Kousarrizi and Ghanbari [119] have developed a ML-based technique for BCI design where different methods for artifacts removal from the EEG signal have been used and then classification was performed on the data extracted from wavelet transform using neural networks and SVM. Murugesan and Sukanesh [120] have proposed an automatic detection method for brain tumor using artificial neural network where a feed-forward neural network was used and trained with BP. They have also used adaptive filtering method for removal of artifacts from the recorded EEG signal. After that, spectral analysis using FFT has been performed for feature extraction. At last, classification was carried out using a back-propagated feed-forward neural network. Jia [121] used artificial neural network and probabilistic neural network for classification of EEG signal. Before that the denoising of signal was performed using time-domain regression method and features have been extracted using autoregressive (AR) model coefficients. Alzoubi et al. [122] performed a test of mental task using EEG data for BCI in on-line and off-line scenario where correlation-based feature selection was used to extract features from EEG data. Before this preprocessing has been performed using common spatial patterns. Then they performed classification task using 13 different classifiers such as ZeroR, 1R, Decision Tree, *k*-Nearest Neighbor, Naive Bayes, RBF Network, SVM, Logistic Regression (LR), Ada Boost, Bagging, Stacking, and Random Forest. Skinner et al. [123] have proposed genetic-based ML classifier for EEG signal classification. Feature extractions have been performed using AR models and FFT where they proposed a genetic-based learning classifier system called as XCS that achieved a maximum accuracy of 99.3% and an average accuracy of 88.9%. Liang et al. [124] have performed a classification of mental tasks using EEG signal with the efficiency of extreme learning machine. They compared the performance of ELM with back-propagation neural network as well as support vector machines. The conclusion derived through this research was that the efficiency of all the techniques is almost same but the training time can be effectively reduced in case of ELM compared to SVM and BPNN. Ioannides et al. [125] have done classification of EEG signal using neural networks. The features have been extracted using wavelet transform technique. Garrett et al. [126] have done an extensive comparison between linear and nonlinear methods for EEG signal classification in BCI system design where they applied SVM for EEG signal classification and the results were compared with neural networks and linear discriminant analysis. The properties of linear and nonlinear SVM are being compared. The SVM is also used for binary and multiclass problems. Aris et al. [127] have designed a classifier model using fuzzy C-means clustering. They collected the data in two different conditions that is in relaxed state and in nonrelaxed state. The features have been extracted using data segmentation and linear regression model. Selim et al. [128] have done an extensive review study on different ML methods used for design of BCI system. Different techniques like linear SVM, linear discriminant analysis, Bayesian linear discriminant analysis (BLDA), generalized Anderson's task, and Fisher linear discriminant analysis were compared. Among all these techniques they found BLDA and SVM have outperformed for all three subjects. Guler et al. [129] have proposed a classifier model based on multiclass support vector machine for classification of EEG signal. Before that, they had also extracted the hidden features from the signal using wavelet transform and Lyapunov exponent. Also, the performances are compared with other benchmark techniques such as multilayer perceptron neural network and probabilistic neural network. Another important task performed in their research is the evaluation of performance of multiclass SVM with error-correcting output codes. Toma-sevic et al. [130] had performed classification of EEG signal for human-computer interaction using neural network trained with back-propagation algorithm. Initially, the EEG denoising has been done using time-domain regression method and AR model was used for feature extraction. They also used probabilistic neural network for classification. Lee and Choi [131] have proposed two methods for classification of EEG signal. One is the combi-nation of PCA and hidden Markov model (HMM). The other one is com-bination of PCA, HMM along with SVM. They also proved the high efficiency of these hybridized techniques. Other than that, they have described the usefulness of PCA in the feature construction phase. Garrett et al. [132] have performed another extensive research on different linear and nonlinear EEG classification methods along with different feature selec-tion methods where a linear system was used, that is, linear discriminant analysis for feature selection and two nonlinear classification techniques such as neural network and SVMs. They presented a feature selection method based on genetic algorithm in application to EEG during finger movement.

2.5 MACHINE LEARNING METHODS FOR EPILEPSY CLASSIFICATION

This section is the discussion about our specific goal that is the application of different ML techniques for epilepsy classification. This process is also some-times referred to as epileptic seizure identification or classification of epilep-tic and NES, and so on. All are pointing to same. The main aim of this process is to distinguish epileptic seizures from nonepileptic ones. Before this we have to first understand what epileptic and NES are. Seizures are nor-mally referred to as the attacks happening in a human brain. Due to this attack there is an abnormal electrical signal flow occurring and the behavior of the person becomes abnormal. Through medical research it has been found that this abnormality may occur due to several reasons. This is most important to identify the reason and origin of this seizure. Epileptic seizures are generally originated from the brain. These are very dangerous attacks where chances of life risk are more. But the NES are not originated from a human brain. Generally, they occur due to any external disturbances in human body such as high blood pressure or high blood sugar. The most dif-ficult task is to differentiate among epileptic and NES because of their similar characteristics or properties. Our mind controls the way we imagine, move, and feel, by passing electrical messages from one brain cell to another. If these messages are disrupted, or too many messages are beamed at once, this causes an epileptic seizure. Approximately one in five people (20%) diagnosed with epilepsy who are then assessed at specialized epilepsy centers are found to possess NES. This may be partly because epilepsy and NES can look really similar and can strike people in similar ways. Nevertheless, the difference between epileptic and NES is their reason. NES are not done by disrupting electrical activity in the brain and so are different from epilepsy (i.e., may be due to low blood sugar, psychological problem, etc.). Nonepileptic seizures are of two types—organic and psychogenic NES. The NES caused due to any physical activity in body is called as organic NES. Whereas psychogenic NES has a neurological cause means they are caused by any mental or emotional process. Epileptic seizures are also divided into two primary types: focal seizures (also called partial seizures) and generalized seizures. Epileptic seizures always start in the brain. The brain has two sides called hemispheres. Each hemisphere has four sections called lobes. Each lobe is responsible for different things such as sight, words, and emotions. If a part of the brain is affected due to seizure then it is called as focal seizure. It may sometimes affect a large part of one hemisphere or a small area in one lobe. Hence, the focal seizure can be simple or complex. Generalized seizures can affect both sides of the brain. This is a very dangerous attack. The person may be unconscious within a few seconds. The treatment of these seizures must be more efficient to overcome the problems arising due to these seizures. Majority of epileptic seizures are handled by some drug therapy. The type of treatment provided again depends on several aspects of the signal such as frequency, severity of the seizures, person's medical history, age, health, and so on. For better treatment to recover from this seizure, it is more important to detect the epileptic seizures. Lot of efforts done by researchers to detect epileptic seizures in human brain through a wide variety of techniques. This process basically includes three different stages. The first stage is to preprocess the raw EEG data in order to remove artifacts or any noise in the signal. After this the second stage is to extract the hidden features from the signal using some feature extraction technique. The last stage is to classify this EEG data using some efficient classification technique to detect epileptic seizures. Guo et al. [133] have done research on automatic detection of epileptic seizures using an ANN where line length features based on wavelet transform multiresolution decomposition were used. For their research work, they considered a publicly available dataset. This dataset was constituted with five different sets from five different persons at different situations where all the five sets were considered with three different combinations such as A-E,

ACD-E, and ABCD-E. A comparison of their work with other techniques was already available. Nigam and Graupe [134] have also done classification of EEG signal for epilepsy identification but only for set A and E using Diagnostic Neural Network along with nonlinear preprocessing filter for feature extraction. Srinivasan et al. [135] also performed classification of datasets A and E by using recurrent neural network with the help of time and frequency features. Kannathal et al. [136] have performed a classification of EEG signal for epileptic seizure identification using adaptive neuro-fuzzy inference system. An entropy measure was taken as features extracted from EEG signal, where they have also classified set A and E [137]. Polat and Günes [138] have performed a feature extraction process for EEG signal using FFT and after that classification of set A and B is carried out by using decision tree technique. Subasi [139] have performed an extensive research and proposed an efficient technique for classification using mixture of expert model. Also, they have extracted the feature using DWT. They also used only two sets A and B. Ocak [140] has performed a binary classification task on two sets. One is combination of A, C, and D and the other set is E. DWT was used for feature extraction and MLPNN for classification. Yuan et al. [141] have performed an extensive study on different kernel-based ML techniques for epilepsy identification through EEG signal classification. The different kernel machines studied include linear SVM, Lagrangian SVM, smooth SVM, proximal SVM, and relevance vector machines. They adopted the feature extraction techniques such as DWT and Lyapunov exponents and have also compared the performance of different mother wavelet functions such as Daubechies wavelet order 2 and 4, Haar wavelet, Symlet wavelet of order 2, and so on. Anusha et al. [142] have used ANN with sliding window technique for pattern classification in EEG signal for epilepsy identification. They used three classes of feature such as time domain, frequency domain, and derived features. Prince and Hemamalini [143] have performed an extensive survey on different soft computing techniques for epileptic seizure identification where they have extracted features using different techniques such as relative intensity ration, Hjorth parameters, fractal dimension, and so on. After this different classification methods are compared such as methods based on neural network and fuzzy logic and based on probabilistic reasoning. Subasi and Ercelebi [144] proposed a classification technique using neural network and logistic regression. Also, they have used a lifting-based DWT for feature extraction from EEG data for epilepsy. They compared the new ML-based approach with statistical model for classification and also used two different training algorithms that is back–propagation and L-M algorithm. Mirowski et al. [145] have done a comparison between SVM and convolutional networks for epilepsy identification using EEG signal classification. The performance is also being compared with a statistical modeling technique, that is, LR. Quirago and Schurmann [146] have proposed another efficient classification of EEG signal using PCA, ICA, LDA, and SVM where they decomposed the signal using DWT for extracting different statistical features. PCA, ICA, and LDA have been used for reducing dimension of dataset and then SVM is applied for classification of epileptic seizures from EEG signal. In this research work, the accuracy of classification techniques was compared with different dimensionality reduction techniques. Mirowski et al. [147] have proposed a classification of patterns of EEG signal for seizure identification where they computed bivariate features from EEG signal such as cross-correlation, nonlinear interdependence, wavelet synchrony, dynamic entertainment, and so on. They used the Freiburg dataset for experimental evaluations. Classification task has been performed by the use of SVM, LR, and convolutional NN.

2.6 SUMMARY

This work mainly focuses on the survey of different research works performed in different aspects of EEG signal analysis. Basically, there are few stages in EEG signal analysis that includes the first step as collection of raw EEG data from a human brain through some recording methodologies. Second, preprocess the EEG data for reduction of unwanted noise or artifacts. These noises or artifacts can be integrated inside EEG due to several reasons. For all type of possibilities the solution must be derived to get the original noiseless signals for further processing. After that, we will get the preprocessed signal; the third task is to analyze it and extract important and hidden features inside it. There are a huge amount of techniques discovered by researchers which assist in further signal analysis methods. Depending on the type of disease analysis we are interested in, the appropriate feature extraction technique can be incorporated to get maximum accurate results. In this research work, we are mainly interested in classification of EEG signal for epilepsy identification. Epilepsy is a neurological disorder disease and it is very much important to distinguish between epileptic and NES for further successful treatment of patients. There are several tasks that can be performed on this EEG to solve any neurological disorder disease or assist in designing a BCI system. Lots of feature extraction techniques are proposed by researchers to enhance the performance of classifier in classification task. From this survey, we concluded that DWT is most efficient and suitable for feature extraction from EEG in case of epilepsy identification. After this signal analysis and feature extraction, the fourth and final task is to design a well-structured and well-defined classifier model to detect any disease or detect any pattern in the signal. This classification task can be performed using several methods which include statistical modeling methods or ML methods. We have also done a survey on comparison of statistical methods and ML methods to classify EEG signal. We found as a conclusion that the ML methods are more efficient and more suitable for this EEG signal analysis to perform different tasks. In this work, our main focus is on classification of EEG signal for epilepsy identification using ML techniques. In the next chapter, an empirical study based on classifiers is performed for the EEG signal classification.

CHAPTER 3

Empirical Study on the Performance of the Classifiers in EEG Classification

Due to the increase in volume and complexity of data, we call for a more efficient method to elicit knowledge from these data. ML techniques are able enough to manage this high volume and complex data. This flood of information requires an automated procedure for data analysis, which is provided by ML techniques. ML is a set of computerized techniques, which focus to automatically learn to discern complex patterns and make sound conclusions based on information. ML has proven its power to reveal hidden information present in large complex datasets. Using ML, it is possible to cluster similar data, classify, or to find an association between various features [148,149]. In the context of EEG signal analysis, ML is the application of algorithms to extract patterns from EEG signals [150]. Still, in that respect are other steps that are likewise carried out, for example, data cleaning and preprocessing, data reduction and projection, incorporation of prior knowledge, proper validation and interpretation of results while analyzing EEG signals. EEG analysis has a number of challenges which make it suited for ML techniques [151].

- EEG comes in large databases.
- EEG recordings are very noisy.
- EEG signals have large temporal variance.

ML is a field broadly evolved from the study of Artificial Intelligence that aims at mimicking intelligent behaviors of human through machines. In ML the most important question is how to make the machine learn. Here, learning is understood in the context of inductive inference. This technique is generally classed into two categories such as supervised learning or predictive learning approach and unsupervised learning or descriptive learning approach. In supervised learning, machine first learns from some labeled data or training information. Then they are made capable to solve unknown examples. But in unsupervised learning one mainly tries to expose the hidden regularities or detect anomalies in data. More or less of the popular ML approaches are neural networks, evolutionary algorithms, fuzzy theory, and probabilistic learning. In this analytical work, our focus is restricted with neural networks and other variants of neural networks for classification of EEG signals such as RBFNN, PNN, RNN, and SVM.

3.1 MULTILAYER PERCEPTRON NEURAL NETWORK

ANN simulates the operation of a neural network structure of the human brain and solves a problem. Like the structure of neurons and connections among them are present in a human brain, the similar structure can be implemented in designing neural network to solve computational problems. A neuron in a brain plays a vital role for information passing, learning patterns, and so on. Similarly, an artificial neuron can be simulated to learn the patterns through communication with other neurons. This is a type of ML technique because it adapts to the environment without explicit programming. It can be able to recognize new patterns by training or learning the known patterns. Various types of complex pattern recognition tasks can be easily solved with the help of ANN. Generally, ANNs are called as perceptron neural network where the term perceptron refers to a model or machine that is created in such a way that it can simulate the behavior of a human brain to identify or recognize. General neural networks are sometimes also referred to as feed-forward neural network because of the absence of a directed cycle, which may be present in some other type of neural network such as RNN. There can be two types of perceptron neural networks such as single-layer perceptron and multilayer perceptron neural networks. The general structure of a neural network constitutes an input layer, an output layer, and one or more hidden layers among them. Each layer consists of a number of neurons connected to other neurons in the next level. Input neurons are generally, single-layer perceptron neural networks that are sufficient for solving linear problems, but nowadays the most commonly employed technique for solving nonlinear problems is MLPNN [15]. The number of layers has been decided based upon number of hidden layers in a network. In a single-layer network there is only one hidden layer along with input and output layer. It can hold various layers such as one input and one output layer along with at least one hidden layer. There are connections between different layers for data transmission. The connections are generally weighted edges to add some extra information to the data and it can be propagated through different activation functions.

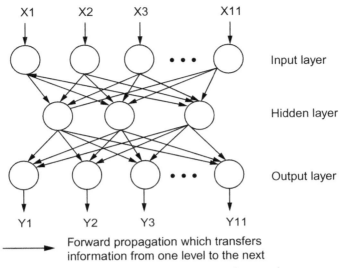

Fig. 3.1 Architecture of a single-layer perceptron neural network.

Fig. 3.1 shows the architecture of a single-layer perceptron neural network and Fig. 3.2 shows the simple architecture of a MLPNN, where x_i denotes the input for the network. Each hidden layer consists of a set of neuron that accept data from its previous layer, after adding some weight to it then passes through a nonlinear activation function. The first hidden layer

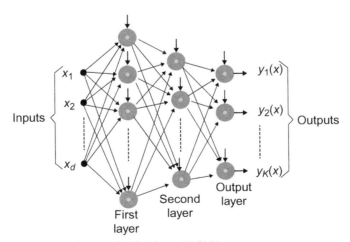

Fig. 3.2 Architecture of a two hidden layer MLPNN.

accepts data from input layer and so on continues. The output layer accepts output from the last hidden layer, also adds some weight and passes through a nonlinear activation function to produce the target output. The output of any perceptron can be represented by following Eq. (3.1).

$$y_i\left(x^{(j)}\right) = \varphi\left(\sum_{k=1}^{n} w_{ik} x_k^{(j)} + b_i\right) \tag{3.1}$$

where

$\varphi(x)$ is the nonlinear activation function

y_i is the output of ith neuron

$x^{(j)}$ is the input of jth layer

$x_k^{(j)}$ is the value of kth neuron in jth layer

w_{ik} is the weight between ith neuron to kth neuron

b_i is the bias value for ith neuron

One of the most important tasks in designing a MLPNN is the selection of an appropriate activation function. There are several functions applied such as Heaviside or step function, sigmoid function, softmax function, Hyperbolic tan function. Eqs. (3.2), (3.3), (3.4), (3.5) describe about these functions, respectively.

$$\varphi(x) = \begin{cases} 0, & x < 0 \\ 1, & x > 0 \end{cases} \tag{3.2}$$

$$\varphi(x) = \frac{1}{1 + e^{-x}} \tag{3.3}$$

$$\varphi(x) = \frac{e^x}{\sum_{j=1}^{n} e^{x_j}} \tag{3.4}$$

$$\varphi(x) = \frac{e^x - e^{-x}}{e^x + e^{-x}} \tag{3.5}$$

The heart of designing an MLPNN is the training of network for learning the behavior of input-output patterns. In this work, we have designed an MLPNN with the assistance of a Java-based environment. This network is developed with the assistance of three popular training algorithms such as BP [16], RPROP [17], and MUR.

3.1.1 MLPNN With Back-Propagation

It is the most widely used propagation training algorithm for feed-forward neural networks [144]. This algorithm is based on gradient descent approach. Hence, it is a requirement for BP that we have to consider an activation function which must be differentiable. Hence, the step function cannot be used for this training algorithm. The gradient descent approach is based on the calculation of change in weight using the gradient value of the error. Error can be formulated as the difference between actual output and network output or target output as shown in Eq. (3.6).

$$E = (o - t) \tag{3.6}$$

where, o — actual output and t — target ouput

This algorithm is different from other algorithms in terms of the weight updating strategies. In BP, generally weight is updated by Eq. (3.7).

$$w_{ij}(k+1) = w_{ij}(k) + \Delta w_{ij}(k) \tag{3.7}$$

where, $w_{ij}(k+1)$ is the weight in next iteration, $w_{ij}(k)$ is the weight in current iteration, and $\Delta w_{ij}(k)$ is the change in weight. The change in weight can be calculated using Eq. (3.8).

$$\Delta w_{ij}(k) = -\eta \frac{\partial E}{\partial w_{ij}}(k) \tag{3.8}$$

where, η is the learning rate chosen in between 0 and 1. With a momentum term the previous equation can be modified as given in Eq. (3.9).

$$\Delta w_{ij}(k) = -\eta \frac{\partial E}{\partial w_{ij}}(k) + \mu \Delta w_{ij}(k-1) \tag{3.9}$$

where μ is the momentum coefficient.

3.1.2 MLPNN With Resilient Propagation

RPROP, short form for resilient propagation [17] is a supervised training algorithm for feed-forward neural network introduced by M. Riedmiller in 1993. It is a type of supervised learning technique similar to BP technique. Instead of magnitude, it takes into account only the sign of the partial derivative or gradient decent and acts independently of each weight. The advantage of RPROP algorithm is that it needs no setting of parameters before applying it. The weight updating is done according to the following equations. Eq. (3.7) is same for the RPROP for weight update.

$$\Delta w_{ij}(k) = \begin{cases} + \Delta_{ij}, \text{ if } \dfrac{\partial E}{\partial w_{ij}}(k) > 0 \\ - \Delta_{ij}, \text{ if } \dfrac{\partial E}{\partial w_{ij}}(k) < 0 \\ 0, \text{ Otherwise} \end{cases} \tag{3.10}$$

$$\Delta_{ij}(k) = \begin{cases} \eta_+ * \Delta_{ij}(k-1), & S_{ij} > 0 \\ \eta_- * \Delta_{ij}(k-1), & S_{ij} < 0 \\ \Delta_{ij}(k-1), & \text{Otherwise} \end{cases} \tag{3.11}$$

where, $S_{ij} = \frac{\partial E}{\partial w_{ij}}(k-1) * \frac{\partial E}{\partial w_{ij}}(k)$, $\eta_+ = 1.2$ and $\eta_- = 0.5$.

The basic idea behind this technique is that it always checks the sign of gradient in current iteration with the sign of gradient in previous iteration. If there is a change in sign then it can be concluded that the technique has produced a local minima and there is a huge change in parameters in previous iteration. Once the choice is made depending on the sign of gradients, the change in weight is calculated which is a very small increase or decrease in the magnitude of previous weights. In this domain of EEG signal classification for epilepsy identification, it has been found that RPROP performs best as compared to all other training algorithms.

3.1.3 MLPNN With Manhattan Update Rule

The basic problem with the BP training algorithm is to ascertain the level to which weights are varied. The MUR only uses the sign of the gradient and magnitude is discarded. If the magnitude is zero, then no change is made to the weight or threshold value. If the sign is positive, then the weight or threshold value is increased by a specific amount defined by a constant. If the sign is negative, then the weight or the threshold value decreases by a specific amount defined by a constant. This constant must be provided to the training algorithm as a parameter.

3.2 RADIAL BASIS FUNCTION NEURAL NETWORK

RBF neural networks are also a type of feed-forward network trained using a supervised training algorithm. The main advantage of RBF network is that it has only one hidden layer and it uses radial basis function as the activation function. These functions are very powerful in approximation. These types of neural networks have attracted many researchers to successfully implement in different problem domains. The RBF network usually trains much faster than back-propagation networks. This kind of network is less susceptible to problems with nonstationary inputs because of the behavior of radial basis function hidden units. The general formula for the output of RBF network [152] can be represented as follows (as shown in Eq. (3.12)), if we consider the Gaussian function as basis function.

$$y(x) = \sum_{i=1}^{M} w_i e^{\left(\frac{-(\|x - c_i\|)^2}{2\sigma^2}\right)}$$
(3.12)

where, x, $y(x)$, c_i, σ, and M denotes input, output, center, width, and number of basis function centered at c_i, similarly w_i denotes weights. For this work, we have constructed a RBFNN by taking the Gaussian function as the basis function and considering randomized centers and width. Fig. 3.3 shows the basic architecture of RBFNN containing exactly three layers, input, hidden, and output layer.

The input layer transforms data to the hidden neurons containing radial basis activation functions. This function generally computes the distance

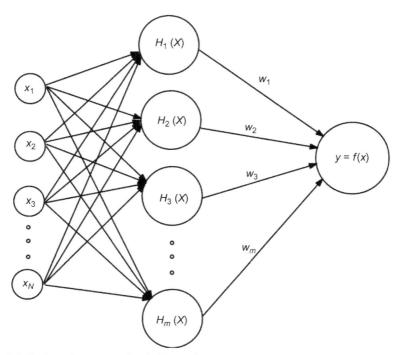

Fig. 3.3 Basic architecture of radial basis function neural network.

between the network inputs and hidden layer centers. The summation of output of hidden layers with some weight is provided as the output of RBFNN. The determination of number of neurons in the hidden layer is one of the most critical and important task because there may be problem of underfitting and overfitting due to number of hidden layer neurons. Insufficient number of neurons may lead to underfitting means the network is not able to recognize patterns properly. Similarly overfitting may lead to poor generalization. Besides Gaussian basis function there are another two radial basis functions used in later research work such as multiquadric and inverse multiquadric basis functions. Eqs. (3.13), (3.14), (3.15) give the formula for all three radial basis functions.

$$\text{Gaussian function}: H_j(x) = \exp\left(\frac{\|x - c_j\|^2}{\sigma^2}\right) \tag{3.13}$$

$$\text{Multiquadric function}: H_j(x) = \sqrt{\left((x - c_j)^2 + \sigma^2\right)} \tag{3.14}$$

$$\text{Inverse multiquadric function}: H_j(x) = \frac{1}{\sqrt{\left((x - c_j)^2 + \sigma^2\right)}} \tag{3.15}$$

Training of the RBFNN can be done in several ways. The most popular method of training a RBFNN is the gradient descent approach as described for BP algorithm. But the training speed of RBFNN is very high as compared to MLPNN with back-propagation. Other than gradient descent there is another method called as Kalman filtering algorithm. Further details about this RBFNN have been discussed in the next chapter.

3.3 PROBABILISTIC NEURAL NETWORK

PNN was first proposed by Specht in 1990. It is a classifier that maps input patterns into a number of class levels. It can be forced into a more general function approximation. This network is organized into a multilayer feedforward network with input layer, pattern layer, summation layer, and the output layer. PNN [22] is an implementation of a statistical algorithm called kernel discriminant analysis (KDA). The advantages of PNN are as follows: it has a faster training process as compared to BP. Also, there are no local minima issues. It has a guaranteed coverage to an optimal classifier as the size

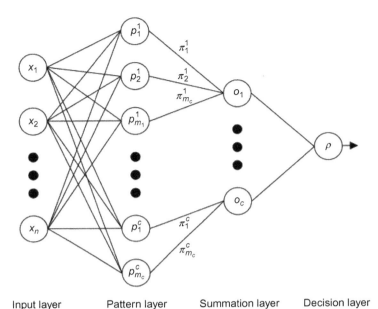

Input layer   Pattern layer   Summation layer   Decision layer

Fig. 3.4 Basic architecture of probabilistic neural network.

of the training set increases. But it has few disadvantages like slow execution of the network because of several layers and heavy memory requirements, and so on. Fig. 3.4 shows the basic architecture of PNN containing four layers.

One of the major differences between PNN and other type of networks is that it defines functions to approximate the probability densities of the sample data rather than fitting the actual data. After the input layer the first layer in the network is the pattern layer. This layer has nodes corresponding to each input in the sample. It passes the input data by applying some weight and the result is passed through some predefined activation function. After that the data passes through summation layer where each node accepts output from pattern nodes associated with a given class. At last the nodes in the output layer give the decision of classification which is a binary value in general.

In PNN [153] a probability distribution function (PDF) is computed for each population. After getting the PDF value we can say, an unknown sample s belongs to a class p if (as shown in Eq. (3.16)),

$$\mathrm{PDF}_p(s) > \mathrm{PDF}_q(s) \forall p \neq q \tag{3.16}$$

where, $\mathrm{PDF}_k(s)$ is the PDF for class k. Other parameters used are prior probability: h, misclassification cost: c, so the classification decision becomes (as shown in Eq. (3.17)),

$$h_p c_p \mathrm{PDF}_p(s) > h_q c_q \mathrm{PDF}_q(s) \forall p \neq q \qquad (3.17)$$

PDF for a single sample can be calculated by using the formula (as shown in Eq. (3.18)),

$$\mathrm{PDF}_k(s) = \frac{1}{\sigma} W\left(\frac{s - s_k}{\sigma}\right) \qquad (3.18)$$

where s—input (unknown), s_k—kth sample, W—weighting function, σ—smoothing parameter. PDF for a single population can be calculated by taking the average of PDF of n samples (as shown in Eq. (3.19)).

$$\mathrm{PDF}_k^n(s) = \frac{1}{n\sigma} \sum_1^n W\left(\frac{s - s_k}{\sigma}\right) \qquad (3.19)$$

The training set in PNN must be a good representative of the total population for getting maximum efficiency of classifier. As the number of sample in training set grows the PNN converges to the Bayes optimal classifier. The training of PNN is nothing but getting the appropriate value of sigma, the smoothing parameter. From the result table, it has been experimentally proved that for epilepsy identification in EEG signal, PNN gives the most accurate result by taking minimum amount of time.

3.4 RECURRENT NEURAL NETWORK

RNN [18] is a special type of ANN having a fundamental feature, that is, the network contains at least one feedback connection [19], so that activation can flow round in a loop. This feature enables the network to do temporal processing and learn the patterns.

Fig. 3.5 shows the basic and simple architecture of a RNN. RNN models can be implemented using several varieties of networks such as,

o   Elman Network
o   Jordan Network
o   Hopfield Network
o   Liquid State Machines
o   Echo State Networks
o   Topology and Weight Evolving Neural Networks

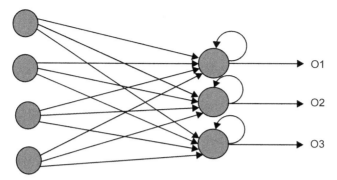

Fig. 3.5 Basic architecture of recurrent neural network.

RNN can be trained using supervised training algorithms such as back-propagation through time (BPTT), real time recurrent learning (RTRL), and extended Kalman filtering (EKF). BPTT is just an extension of back-propagation training algorithm used in MLPNN. BP training is not suitable for RNN because it assumes there is no cycle in the network. Hence, BPTT has been introduced that unfolds the RNN into a group of feed-forward neural network in time, by making a stack of identical copies of RNN. Similarly, RTRL is also a gradient descent method-based training algorithm that calculates the error gradient at each time step. EKF is derived from the Kalman filtering method for nonlinear systems. The most important common features shared by all types of RNN [20,21] are as follows: they incorporate some form of multilayer perceptron as subsystem; they implement the nonlinear capability of MLPNN [22,23] with some form of memory. In this research work the ANN architecture has been used and implemented for modeling and classifying the Elman recurrent neural network (ERNN). It was originally developed by Jeffrey Elman in 1990. The BPTT learning algorithm is used for training [154,155], which is an extension of BP that performs gradient decent on a complete unfolded network. If a network training sequence starts on time t_0 and ends at time t_1, the total cost function can be calculated according to Eq. (3.20):

$$E_{total}(t_0, t_1) = \sum_{t=t_0}^{t_1} E_{sse}(t) \quad ce \quad (3.20)$$

And the gradient decent weight update can be calculated as (shown in Eq. (3.21)):

$$\triangle w_{ij} = -\eta \sum_{t=t_0}^{t_1} \frac{\overline{\partial E_{sse}(t)}}{ce}}{\partial w_{ij}} \tag{3.21}$$

For this research work we have considered the RPROP training as the primary training algorithm. A secondary training algorithm has been used for improving the performance of RNN, that is, Simulated Annealing. The different parameters and their values used in this work have been discussed in the next section.

3.5 SUPPORT VECTOR MACHINES

SVM is the most widely used ML technique based on pattern classification nowadays. It is based on statistical learning theory and was developed by Vapnik in 1995. Statistical learning methodology generally produces to a platform for studying the problem to retrieve information, performing predictions, and taking decisions. It also helps in selecting a suitable hyperplane to correctly match the underlying function. The primary aim of this technique is to project nonlinear separable samples onto another higher dimensional space by using different types of kernel functions. In late years, kernel methods have received major attention, especially due to the increased popularity of SVMs [24]. Kernel functions play a significant role in SVM [25] to bridge from linearity to nonlinearity. Least square SVM [26] is also an important SVM technique that can be applied for classification task [27]. Extreme Learning Machine and Fuzzy SVM [28–30], and Genetic algorithm tuned expert model [31] can also be applied for the purpose of classification. Fig. 3.6 shows the graph for creating a linear separable line to classify a simple binary classification problem using SVM.

In this analytical work, we have evaluated three different types of kernel functions [32], that is, Linear, Polynomial, and RBF kernel [32]. Kernel methods can be defined as a group of algorithms for pattern recognition basically used in SVMs. The basic task of a kernel is to investigate general types of relations or patterns. This method normally maps the nonlinear separable data into a high dimensional space in a hope that it be linearly separable. These mapping functions are called as kernel functions. Linear kernel is the simplest kernel function available. Kernel algorithm using a linear kernel is often equivalent to their nonkernel counterparts [41]. It can be represented by Eq. (3.22).

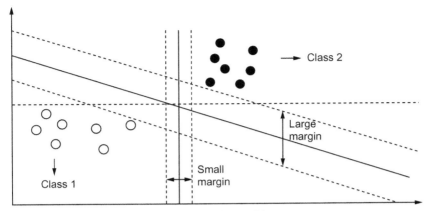

Fig. 3.6 Linear SVM classification for two-class problem.

$$K(x, y) = x^T y + c \tag{3.22}$$

From the result table, it can be clearly understood that for a classification problem consisting of only sets A & E or D & E, it is providing 100% accuracy. But it is not able to classify properly for sets A + D & E.

Polynomial kernel is a nonstationary kernel. This kernel function can be represented as given in Eq. (3.23).

$$K(x, y) = \left(\alpha x^T y + c \right)^d \tag{3.23}$$

where, α, c, and d denote slope, any constant, and degree of polynomial, respectively. Generally, the degree is taken as 2 because higher degree of polynomial may sometimes lead to the problem of overfitting.

Somehow this kernel function [156,157] is better as compared to linear kernel function. However, the RBF kernel function [158] has been proven as the best kernel function used for this application, which can classify different groups with 100% accuracy with a minimum time interval. Eq. (3.24) represents the RBF kernel also called as Gaussian kernel.

$$K(x, y) = e^{-\frac{\|x-y\|^2}{2\sigma^2}} \tag{3.24}$$

Here, the parameter σ plays an important role which is adjustable according to the requirements for the kernel. The correct estimation of this value is important because overestimation and underestimation both have their own disadvantages with respect to specific problems. The major strength of SVM is the simplicity of training methodology. There is no scope for local optimum values like other techniques (Neural Networks). Its automatic mapping feature to a high dimensional space makes it easier for problem data analysis.

3.6 EXPERIMENTAL STUDY

This section gives an experimental evaluation of the empirical study on different classification techniques based on ML approach for detection of epilepsy in EEG brain signal through a set of proper experimental evaluations. Various experiments are done to validate this empirical study. The ML-based classifiers are proved as the most efficient way for pattern recognition. It aids to design models that can learn from some previous experience (known as training) and further it can be used to recognize appropriate patterns for unknown samples (known as testing).

3.6.1 Datasets and Environment

In the present work, we have collected data from Ref. [37] which is a publicly available database related to diagnosis of epilepsy. This resource provides five sets of EEG signals. Each set contains reading of 100 single-channel EEG segments of 23.6 s duration each. These five sets are described as follows. Datasets A and B are considered from five healthy subjects using a standardized electrode placement system. Set A contains signals from subjects in a slowed down state with eyes open. Set B also contains signal same as A but ones with the eyes closed. The datasets C, D, and E are recorded from epileptic subjects through intracranial electrodes for interictal and ictal epileptic activities. Set D contains segments recorded from within the epileptogenic zone during seizure-free interval. Set C also contains segments recorded during a seizure-free interval from the hippocampal formation of the opposite hemisphere of the brain. Set E only contains segments that are recorded during seizure activity. All signals are recorded through the 128-channel amplifier system. Each set contains 100 single-channel EEG data. In all there are 500 different single-channel EEG data. In the next section we will illustrate how to crack these signals using discrete wavelet transform [9] and prepare several statistical features and form a proper sample feature dataset. For this research experiments we have considered three different cases of dataset depending on three different combinations of sets available. In the first case, we have considered set A as one class and set E as other class. In the second case, we have considered set D as one class and set E as other class. In the last and third case, we have taken the collection of set A and D combined together as one class and set E as another class.

All experiments for this research work are performed using a powerful Java Framework known as Encog [159] developed by Jeff Heaton and his team. Currently, we are using Encog 3.2 Java framework for all experimental result evaluation. This is the latest version and it supports almost all the features of

ML techniques. Along with this framework, there are a lot of packages, classes, and methods that have been defined to support the experimental evaluations. Java is the most potent and efficient language nowadays. The rightness of the experimental works can be verified easily using this language. There are almost nine different ML algorithms that have been implemented for EEG signal classification for epileptic seizure detection.

3.6.2 Parameters

The Encog Java framework provides a vast circle of library classes, interfaces, and methods that can be utilized for designing different ML-based classifier models. There are lists of parameters (as presented in Table 3.1) required to be set for smooth execution of models.

3.6.3 Results and Analysis

Here, we have discussed about the performance of all ML-based classifiers for classifying EEG signal. The different measures used for performance estimation are like specificity (SPE), sensitivity (SEN), accuracy (ACC), and time elapsed for execution of models. These estimates can be derived from the following confusion matrix given in Table 3.2.

So the formula for different estimation measures is given in Eqs. (3.25), (3.26), (3.27).

$$\text{Accuracy} = \frac{\text{TN} + \text{TP}}{\text{TN} + \text{TP} + \text{FN} + \text{FP}} \qquad (3.25)$$

$$\text{Sensitivity} = \frac{\text{TP}}{\text{TP} + \text{FN}} \qquad (3.26)$$

$$\text{Specificity} = \frac{\text{TN}}{\text{TN} + \text{FP}} \qquad (3.27)$$

From the evaluation result given in Table 3.3 it is clear that MLPNN with RPROP is the most efficient training algorithm both in conditions of accuracy as well as the amount of time needed to execute the programs requiring different shapes such as A&E, D&E, and A + D & E. This MLPNN technique can be compared with other ML techniques.

Table 3.4 presents a comparison of different Kernel types used for classification using SVM. It is the most powerful and efficient ML tool for designing classifier model. This table clearly shows quite a good result for SVM with RBF kernel.

Table 3.5 defines a list of experiments led by studying different forms of Neural Network, such as RBFNN, PNN, and RNN. It suggests the

Table 3.1 List of parameters for models execution

Classification techniques	Required parameters and values
MLPNN/BP	Activation function—sigmoid Learning rate = 0.7 Momentum coefficient = 0.8 Input bias—yes
MLPNN/RPROP	Activation function—sigmoid Learning rate = NA Momentum coefficient = NA Input bias—yes
MLPNN/MUR	Activation function—sigmoid Learning rate = 0.001 Momentum coefficient = NA Input bias—yes
SVM/linear	Kernel type—linear Penalty factor = 1.0
SVM/polynomial	Kernel type—polynomial Penalty factor = 1.0
SVM/RBF	Kernel type—radial basis function Penalty factor = 1.0
PNN	Kernel type—Gaussian Sigma low—0.0001 (smoothing parameter) Sigma high—10.0 (smoothing parameter) Number of sigma—10
RNN	Pattern type—Elman Primary training type—resilient propagation Secondary training type—simulated annealing *Parameters for SA* Start temperature—10.0 Stop temperature—2.0 Number of cycles—100
RBFNN	Basis function—inverse multiquadric Center and spread selection—random Training type—singular value decomposition

Table 3.2 Confusion matrix

		Predicted class	
		Class 1	Class 2
Actual class	Class 1	TP	FN
	Class 2	FP	TN

effectiveness of using PNN for classification of EEG signal for detecting epileptic seizures.

Table 3.6 gives a detailed empirical analysis of the performance of different classification techniques based on ML approaches.

Table 3.3 Experimental evaluation result of MLPNN with different training algorithms

Multilayer perceptron neural network with different propagation training algorithms

Cases for seizure types	Back-propagation				Resilient-propagation				Manhattan update rule			
	SPE	SEN	ACC	TIME	SPE	SEN	ACC	TIME	SPE	SEN	ACC	TIME
Case 1 (A,E)	100	90.09	94.5	16.52	99.009	100	99.5	2.846	97.29	77.77	85	7.541
Case 2 (D,E)	100	83.33	90	22.22	99.009	100	99.5	2.547	55.68	78.78	60	7.181
Case 3 (A+D, E)	100	86.95	92.5	23.12	95.85	85.98	92.33	14.79	93.78	82.24	89.66	14.85

Table 3.4 Experimental evaluation result of SVM with different kernel types

Cases for seizure types	Support vector machine with different kernel types											
	Linear				Polynomial				RBF			
	SPE	SEN	ACC	TIME	SPE	SEN	ACC	TIME	SPE	SEN	ACC	TIME
Case 1 (A,E)	100	100	100	2.127	100	100	100	2.101	100	100	100	2.002
Case 2 (D,E)	100	100	100	1.904	100	100	100	1.902	100	100	100	2.021
Case 3 (A+D, E)	90.67	76.63	85.66	11.61	100	99.009	99.66	7.24	100	100	100	2.511

Table 3.5 Experimental evaluation result of RBFNN, RNN, PNN with different training algorithms

Cases for seizure types	RBF neural network				Probabilistic neural network Other types of neural network				Recurrent neural network			
	SPE	SEN	ACC	TIME	SPE	SEN	ACC	TIME	SPE	SEN	ACC	TIME
Case 1 (A,E)	83.076	65.925	71.5	2.051	100	100	100	0.967	77.173	73.148	75	10.31
Case 2 (D,E)	100	97.08	98.5	1.828	100	100	100	0.977	64.705	71.604	67.5	13.29
Case 3 (A +D,E)	92.30	66.41	81	2.928	100	100	100	1.616	67.346	66.666	67.333	19.58

Table 3.6 Comparative analysis of different machine learning classification techniques

Machine learning classification technique	Case 1 (set A & E)		Case 1 (set D & E)		Case 1 (set A+D & E)	
	Overall accuracy in % age	Approximate time taken in seconds	Overall accuracy in % age	Approximate time taken in seconds	Overall accuracy in % age	Approximate time taken in seconds
MLPNN/BP	94.5	16.527	90	22.226	92.5	23.127
MLPNN/RP	99.5	2.846	99.5	2.547	92.33	14.798
MLPNN/MUR	85	7.541	60	7.181	89.66	14.85
SVM/linear	100	2.127	100	1.904	85.66	11.61
SVM/ploy	100	2.101	100	1.902	99.66	7.24
SVM/RBF	100	2.002	100	2.021	100	2.511
PNN	100	0.967	100	0.977	100	1.616
RNN	75	10.31	67.5	13.29	67.33	19.58
RBFNN	71.5	2.051	98.5	1.828	81	2.928

3.7 SUMMARY

The detection of epileptic seizure in EEG signal can be performed by classifying these signals collected from different patients in different situations. This classification can be accomplished by using different ML techniques. In this work, we have compared the functioning and efficiency of different ML techniques like MLPNN, RBFNN, RNN, PNN, and SVM for classification of EEG signal for epilepsy identification. Further, the tool MLPNN uses three training algorithms BP, RPROP, and MUR. Similarly, SVM uses three kernels such as Linear, Polynomial, and RBF kernels. This comparative study clearly shows the difference in the efficiency of different algorithms with respect to the task for classification task. From the previous experimental study, we can conclude that SVM is the most efficient and powerful ML technique for classification purpose. Again, SVM with RBF kernel provides the maximum accuracy of the classification task. PNN is also a good competitor for SVM for this application. But as compared to SVM, PNN has some extra overhead of setting parameters. Lot of research is still going on to modify some algorithms to increase their efficiency by incorporating some optimization algorithms, like we can also enhance the performance of RBFNN to achieve the required accuracy. In the following chapter, a RBF trained improved PSO algorithm is proposed for the identification of epilepsy detection.

CHAPTER 4

EEG Signal Classification Using RBF Neural Network Trained With Improved PSO Algorithm for Epilepsy Identification

EEG [7] is the brain signal generated in brain due to a collision of huge number of neurons among each other. This collision generates very small amount of electrical signal. EEG [7] measures this electrical activity to study the human behavior. Careful analysis of these signals can lead to detection of many disorders. About 1% of the total population in this world are affected by this disease. So, it is necessary to identify and properly diagnose this disease. If a person has a seizure then it does not necessarily mean that the person is affected by epilepsy [160]. So, it is very difficult to recognize and differentiate between epileptic seizures and others by normal human eyes.

There are several methods available to record the EEG signal. There is a famous 10–20 electrode placement scheme to measure EEG. In this system there are different electrodes placed on the human scalp to record the EEG activity. The electrodes are placed in a 10–20 international standard. Then these electrical activities in a human brain are recorded by machine connected to these electrodes through some wires.

Generally, doctors take a printed copy of these recorded signals and identify whether there is any sign of epilepsy or not. But it is very difficult to differentiate among normal seizures and epileptic seizure through normal eyes. Hence, it is required to develop such a system in which we can analyze the EEG signal [4] and properly differentiate between normal and epileptic seizure [161].

In this research work, we have taken the help of MATLAB to analyze the EEG signals. According to the literature survey, it is clearly understood that the DWT is the most efficient method to analyze the EEG signals. This method generally fits to the problem where the signals are very transient in nature. That is, the frequency of signal changes rapidly with respect to time [162]. After analysis of EEG signals by DWT [140, 163] we can discover several statistical features which can be used for further processing.

After analysis of signal the most important phase is to classify the signals whether they are epileptic seizures or normal. Classification is a fundamental task in data mining. This method is used for identifying a specific data sample as to which prespecified group it belongs to. In this problem, we have specified two groups, one is normal and the other is epileptic seizure group. Classification of seizures in EEG signal is one of the challenging tasks. In classification, we are given a set of instances consisting of several features or attributes called as training set. One of the attribute called the classifying attribute identifies the class to which each instance belongs. Another set of unknown instances called as testing set are used for measuring the efficiency of classifier model.

Over many years ANN have been very widely used in many biomedical signal analyses because they classify the signals efficiently for decision making. Every classification system must be provided with a set of sample data that is represented by features extracted from a signal. Different methods used for this can be frequency domain features, wavelet transform, and so on. Over the years there is several other architecture of NN model [164] that has been used such as MLPNN, ANFIS, RBF, RNN, and FNN. Our classification method is based on RBFNNs. It is popularly used in many research areas because of its features like universal approximation, compact topology, and faster learning speed.

The fundamental constraint in any classification method is its learning procedures. For any ML approach it is very important to select the best learning method for classification. From our previous study [164] we have concluded that RBFNN model requires a better learning procedure for classification of EEG signal. In this work, an improved version of particle swarm optimization (PSO) algorithm is used to train the RBF network for classification of EEG signal for epileptic seizure identification.

4.1 RELATED WORK

There are many researchers, who have proposed a number of methods to increase the performance of RBFNN in different applications. But in the application of EEG signal classification it is a completely new area. Several modified training methods have been proposed such as Fathi and Montazer [165] have proposed a novel PSO-OSD algorithm to improve the RBF learning algorithm in real-time applications. Mazurowski et al. [166] have proposed a method for neural network training and compared with BP algorithm for medical decision making. Zhang et al. [167] have proposed a hybrid PSO-BP algorithm for training feed-forward neural network. Ge et al. [168] proposed a modified PSO algorithm for training recurrent neural network.

Zhao and Yang [169] have proposed a modified PSO algorithm called as CRPSO to train neural network for time series prediction. Guerra and Coelho [170] have proposed a novel method to train RBFNN using PSO and *k*-means clustering technique. From these literature surveys it is clearly understood that there are lot of researches that have been done by researchers for performance enhancement of ANN using PSO algorithm and also variations of PSO.

4.2 RADIAL BASIS FUNCTION NEURAL NETWORK

RBFNN is one of the simplest form of ANN consisting of exactly three layers, namely, input, hidden, and output layer (as shown in Fig. 4.1).

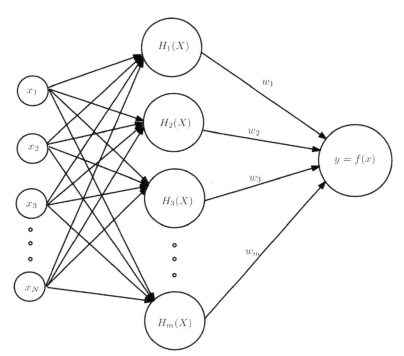

Fig. 4.1 Radial basis function neural network architecture.

The restriction of only three layers makes it simplest and somehow efficient ANN architecture. The idea of RBFNN has been derived from function approximation. An RBF network positions one or more RBF neurons in the space described by the predictor variables. This space has as many dimensions as there are predictor variables. The Euclidean distance is computed from the point being evaluated to the center of each neuron. The RBF is so named because the radius distance is the argument to the function. Output of RBFNN depends on the distance of the input from a given stored vector.

4.2.1 RBFNN Architecture

For this research work, we have taken N number of input neurons, m number of hidden neurons, and one output neuron. There are several kernel functions used in RBFN, such as Gaussian, multiquadric, and inverse multiquadric. Each of the functions has its own benefits depending on the data domain they are used. Based on the recommendation of our previous research, we used to verify the performance of Gaussian, multiquadric, inverse multiquadric basis function in RBFNs for identification of epileptic seizure, but it was found that the performance of inverse multiquadric is pretty higher than the performances of the other two.

The different symbols and dimensions used in the above figure are as follows (as given in Table 4.1).

The above symbols can be further described as follows.

$$\text{Input vector } (X) = \{x_1, x_2, ..., x_N\}$$

$$\text{Hidden neurons} = \{H_1(X), H_2(X), H_3(X), ..., H_m(X)\}$$

$$\text{Weight vector } (W) = \{w_1, w_2, ..., w_m\}$$

Table 4.1 Parameter description for RBFNN

Parameter symbol	Description	Dimension
n	Number of samples vectors	200 or 300
D	Desired output vector	200×1 or 300×1
M	Number of hidden neurons	40
W	Weight vector	40×1
N	Number of input neurons	20
X	Input vector	1×20
C	Center matrix	40×20
σ	Spread vector	40×1

$$\text{Center matrix}\,(C) = \begin{bmatrix} c_{11} & \cdots & c_{1N} \\ \vdots & \ddots & \vdots \\ c_{m1} & \cdots & c_{mN} \end{bmatrix}$$

$$\text{Spread vector}\,(\sigma) = \{\sigma_1, \sigma_2, \ldots, \sigma_m\}$$

Hence, the output of RBFNN can be defined as in Eq. (4.1),

$$\text{RBFNN output}\,(y) = \sum_{j=1}^{m} w_j * H_j(X) \tag{4.1}$$

where $H_j(X)$ can be any one of the function given in Eqs. (4.2), (4.3), (4.4).

$$\text{Gaussian function},\, H_j(X) = \exp\left(\frac{\|X - c_j\|^2}{\sigma^2}\right) \tag{4.2}$$

$$\text{Multiquadric function},\, H_j(X) = \sqrt{\left((X - c_j)^2 + \sigma^2\right)} \tag{4.3}$$

$$\text{Inverse multiquadric function},\, H_j(X) = \frac{1}{\sqrt{\left((X - c_j)^2 + \sigma^2\right)}} \tag{4.4}$$

$$\text{where},\, (X - c_j)^2 = \sum_{k=1}^{N} (x_k - c_{jk})^2 \tag{4.5}$$

To measure the performance of training algorithms, error is calculated by finding the difference between desired output and actual output. Hence, the MSE function can be defined as (given in Eq. 4.6),

$$\text{MSE}(c, \sigma, w) = \frac{1}{n}\left(\sum_{i=1}^{n}\left(d_i - \sum_{j=1}^{m} w_j * H_j(X)\right)\right)^2 \tag{4.6}$$

4.2.2 RBFNN Training Algorithm

Learning or training of a network is a process by which it adapts to the environment by adjusting few parameters. For RBFN, to get the desired output for a given input there are mainly three adjustable parameters, such as center, spread, and weight. There are several learning algorithms proposed by several researchers among which gradient descent approach is the most commonly used. This is a first-order derivative-based optimization algorithm for finding local minimum of a function. According to

Eq. (4.6), the error can be calculated by finding the difference between desired and actual output. Then the partial derivative of this error with respect to weight and center can be calculated to adjust the parameter with minimizing the error. The formula of gradient descent is given as follows (as shown in Eq. 4.7).

$$w_i = w_i - \eta \frac{\partial E}{\partial w_i} \text{ and } c_{ij} = c_{ij} - \eta \frac{\partial E}{\partial c_{ij}} \tag{4.7}$$

where η is the learning parameter or step size. We have performed several experimental evaluations by considering different η values between 0.5 and 1.0. The detailed results are given in the next section. There are also several other learning techniques like PSO [171], differential evolution [172], genetic algorithm (GA) [173], and so on. Basically, RBF networks are used in many applications because of its architectural simplicity and requirement of less number of adjustable parameters. Therefore, to employ the RBFNN in the relevance of EEG classification, we require some supplementary techniques for improving its performance. This can be done by integrating optimization techniques with the training methods. There are several optimization techniques available such as PSO, ABC, and GA. Yet again, we opt for PSO optimization technique owing to its requirement for less number of adjustable parameters and its capability to produce global optimal solutions. In this study, we have proposed a new innovative training algorithm for RBFN-based PSO algorithm. The different parameters such as center, spread, and weight are trained by using PSO optimization algorithm. This is biologically inspired algorithm from the behavior of bird swarms or fish flocks. It has been explained in Section 4.3.

4.3 PARTICLE SWARM OPTIMIZATION

PSO algorithm was developed initially by Kennedy and Eberhart in 1995. This algorithm is a nature inspired algorithm that is inspired from the behavior of bird flocks called as swarm [35]. In this algorithm each solution is represented as a vector called as a particle (bird). It is a population-based algorithm where each solution is treated as a particle in n-dimensional space. The population (swarm) may contain any random number of initial solutions (particles). Each particle starts with its initial position and velocity, and then moves in the solution space to achieve the optimum solution. The main computational steps of PSO include generating initial position and velocity of each particle in population, updating position and velocity for a certain number of generations to get the optimal solution. Let us discuss about the mathematical computation of PSO algorithm. Let any particle $\vec{x_k}$ (solution) in n-dimensional space is represented in Eq. (4.8).

$$\vec{x_k} = \{x_{k1}, x_{k2}, x_{k3}, \ldots, x_{kn}\} \qquad (4.8)$$

where $k = 1, 2, 3, \ldots, d$ and d is the number of particles in the swarm. Each particle maintains its own velocity represented as given in Eq. (4.9).

$$\vec{v_k} = \{v_{k1}, v_{k2}, v, \ldots, v_{kn}\} \qquad (4.9)$$

Also, in this algorithm each particle maintains its personal best position called as p_{best} and a best solution among all the particles called as g_{best}. In each iteration or generation the particles move toward optimal solution by updating their velocity and position according to the formula given in Eqs. (4.10), (4.11).

$$\vec{v_k}(t+1) = \lambda * \vec{v_k}(t) + c_1 * r_1 * \left(\overrightarrow{p_{best\,k}}(t) - \vec{x_k}(t)\right) + \left(\overrightarrow{g_{best\,k}}(t) - \vec{x_k}(t)\right)$$
$$(4.10)$$

$$\vec{x_k}(t+1) = \vec{x_k}(t) + \vec{v_k}(t+1) \qquad (4.11)$$

where $\vec{v_k}(t+1)$ represents the velocity of kth particle at $t+1$ iteration. λ is the inertia weight, $\vec{v_k}(t)$ represents velocity of kth particle at t iteration. $\overrightarrow{p_{best\,k}}(t), \overrightarrow{g_{best\,k}}(t)$ represent the personal best of the particle and global best of swarm at t iteration, respectively. $\vec{x_k}(t), \vec{x_k}(t+1)$ are the previous and present solutions, respectively. c_1 and c_2 are two positive real constants known as *self-confidence factor* and *swarm confidence factor*, respectively. r_1 and r_2 are any random number generated in between [0, 1]. From the survey, it has been proved that larger inertia weight performs more efficient global search and smaller inertia weight performs efficient local search [174]. Hence, this inertia weight can be considered as an important parameter to tune the performance of PSO algorithm. This chapter presents and proposes a novel strategy to vary the inertia weight in each iteration to perform the efficient global search [175].

4.3.1 Architecture

Fig. 4.2 describes the architecture of basic PSO algorithm.

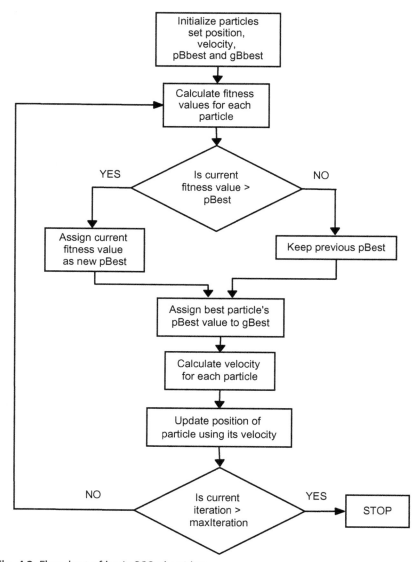

Fig. 4.2 Flowchart of basic PSO algorithm.

4.3.2 Algorithm

Following Algorithm 4.1 gives a brief algorithm for the implementation of basic PSO technique.

Algorithm 4.1 Basic PSO

Step 1: Initialize all the particles position (x), velocity (v), p_{best} and g_{best}

Step 2: Repeat Step 3 to 11 until Iteration < MAX_Iteration

    Step 3: Repeat Step 4 to 7 for each particle (i) in population (S)

        Step 4: if fitness(x_i) < fitness(p_{best})

        Step 5: $p_{best} = x_i$

        Step 6: if fitness(p_{best}) < fitness(g_{best})

        Step 7: $g_{best} = p_{best}$

    Step 8: Repeat Step 9 and 10 for each particle (i) in population (S)

        Step 9: $\vec{v_k}(t+1) = \lambda * \vec{v_k}(t) + c_1 * r_1 * \left(\overrightarrow{p_{best\,k}}(t) - \vec{x_k}(t)\right) + \left(\overrightarrow{g_{best\,k}}(t) - \vec{x_k}(t)\right)$

    Step 10: $\vec{x_k}(t+1) = \vec{x_k}(t) + \vec{v_k}(t+1)$

    Step 11: Iteration ++

Step 12: Stop

4.4 RBFNN WITH IMPROVED PSO ALGORITHM

This section describes the new proposed learning method for RBFNN known as Improved PSO (IPSO) algorithm for classification task. Along with IPSO details, this section describes how it is used for learning RBFNN and finally an algorithmic description about the proposed model. One of the important drawbacks of PSO algorithm is its very slow searching around the global optimum. The IPSO algorithm is based on general PSO algorithm. The main idea of improving the base algorithm is to do faster search around the global optimum [171]. Hence, the basic PSO algorithm has been improved as follows.

In Eq. (4.10) the inertia weight (λ) is generally taken as a constant value for the total number of generations. This can be modified by decreasing λ gradually as the number of generations (or iteration) increases. Thus we can reduce the search space for global optimum by reducing the value as the number of generation increases. After each generation the best particle in the previous generation will replace the worst particle in current generation. Several selection strategies have been proposed by researchers. In this research work, we have applied two types of selection strategy sequentially for inertia weight, one is linear selection and other is nonlinear selection. In linear selection λ should reduce rapidly, while around the optimum λ will reduce slowly. Mathematically, it can be described as follows.

Let λ_0 is the initial value of inertia weight, λ_1 is the end point of linear selection, g_1 is the number of generations for linear selection, and g_2 is the number of generations for nonlinear selection. Then according to the proposed algorithm for 1 to g_1 number of generations the inertia weight for PSO will be calculated as given in Eq. (4.12).

$$\lambda_1 = \lambda_0 - ((\lambda_1/g_1) * i) \tag{4.12}$$

where

$$i = 1, 2, 3, \dots, g_1$$

For g_1 to g_2 number of generations the inertia weight for PSO will be calculated according to Eq. (4.13).

$$\lambda_1 = (\lambda_0 - \lambda_1) * e^{((g_1 + 1) - i)/i} \tag{4.13}$$

where

$$i = g_1, \dots, g_2$$

Generally the values of g_1 and g_2 are selected from empirical study. For this research work, we have considered the total number of generations as 100. Linear and nonlinear selection of inertia weight takes place for 50% of the total number of generations. The detailed experimental evaluation of this process has been explained in the next section.

4.4.1 Architecture of Proposed Model

This section describes the detailed procedure for the proposed model. The classifier model consists of mainly three phases. In the first phase, the data preprocessing is done. Since we are considering EEG signal classification for epilepsy as our problem domain, hence the preprocessing of data is necessary. But EEG data for eye state prediction, the preprocessing is not required. The signal analysis and feature extraction is done by using DWT (Fig. 4.3).

In the second phase some portions of these datasets are provided for training RBFNN using our proposed IPSO algorithm. The detailed algorithmic procedures are described in the next section. In the last and third phase, the network model is tested by using remaining portions of dataset. This testing of classifier model also includes the validation procedure. The different measures to calculate efficiency of model have been described in the next section.

4.4.2 Algorithm for Proposed Model

The following algorithm/pseudo–code (Algorithm 4.2) describes the detailed structure of our proposed model.

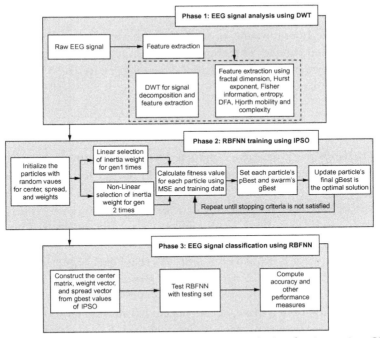

Fig. 4.3 Proposed model architecture for EEG signal classification using RBFNN with IPSO.

Algorithm 4.2 Proposed Model of RBFNN Trained With IPSO Algorithm

For each particle do

    Initialize particle position and velocity.

End For

While stopping criteria is not fulfilled do

    Calculate the inertia weight using Eq. (4.12) or (4.13) depending on generation number.

    For each particle do

        Calculate fitness value (Using MSE of RBFNN).

        If fitness value is better than best fitness value in particle history (p_{best})

            Then *Set current position as p_{best}.*

    End If

End For

Choose the global best (g_{best}) as the particle with best fitness value among all the particles.

For each particle do

    Calculate particle velocity using Eq. (4.10)

    Update particle position (Center, Spread & Weight) using Eq. (4.11)

End For

End While

4.5 EXPERIMENTAL STUDY

This section describes the detailed analysis of experimental works carried out for our proposed model. The computational complexity of the proposed algorithm may vary for different datasets depending on its size. The parametric values may vary accordingly.

4.5.1 Dataset Preparation and Environment

In this research work we have conducted several experiments on two different types of datasets. One of them is EEG dataset for epileptic seizure identification and the other one is EEG dataset for eye state prediction. These are openly available source of data for EEG used by many researchers for their research work. EEG data for epilepsy is mainly categorized into five types: set A, B, C, D, and E. Each set contains 100 single-channel EEG segment. Each segment is of 23.6 s duration. All these data have been prepared by removing artifacts due to eye or muscle movements. Set A and B have been collected from healthy patients having eyes open and closed, respectively. Set C, D, and E have been collected from epileptic patients, but C and D recorded in seizure-free activity, where set E contains seizure activity. EEG data for eye state prediction is already in sample-feature format for classification problem (Table 4.2).

This research work has been supported by a lot of experimental evaluations. These results of evaluation have been used for providing correctness of our proposed model. All the experiments have been carried out in Java platform. The latest version of Java is used, that is, JDK 1.8 with Eclipse Mars as IDE. The operating system used is Linux Mint 17.2 with hardware configuration as RAM of 2GB and Intel processor. There are several tools of Java that has been used such as classes, packages, enumerations, interfaces.

Table 4.2 Description of benchmark EEG dataset for epilepsy identification and eye state prediction

Datasets		No. of features	No. of classes	No. of patterns
EEG dataset for epilepsy identification	Set (A and E)	29	2	200
	Set (D and E)	29	2	200
	Set (A+SD and E)	29	2	300
EEG dataset for eye state prediction		14	2	14,980

Several frameworks have been designed for ML techniques, optimization techniques, graph drawing and designing.

4.5.2 Parameters

The different significant parameters used for RBFNN are center, spread, and weight. The different symbols used for RBFNN, PSO, and IPSO are described in Tables 4.3 and 4.4.

Generally, the evaluation of a classification problem is based on a matrix called as confusion matrix with number of testing samples correctly classified and incorrectly classified. As a result based on that, the accuracy can be measured according to Eq. (4.14).

$$\text{Accuracy} = \frac{\text{TN} + \text{TP}}{\text{TN} + \text{TP} + \text{FN} + \text{FP}} \tag{4.14}$$

Table 4.3 Description of parameters used for RBFNN

Symbols used	Description	Considered value/size
n	Number of input vectors	200 or 300
D	Desired output vector	200×1 or 300×1
M	Number of hidden neurons	40
W	Weight vector	40×1
N	Number of input neurons	29 or 14
X	Input vector	1×29
C	Center matrix	40×29
σ	Spread vector	40×1

Table 4.4 Description of parameters used for IPSO

Symbols used	Description	Considered value/size
λ_0	Initial inertia weight	0.8
λ_1	Final inertia weight for linear selection	0.5
c_1	Local search coefficient	0.9
c_2	Global search coefficient	0.9
P	Population size	25
g_1	Number of generations for linear increment	50
g_2	Number of generations for nonlinear increment	50

For a binary classification problem the other measures include Precision or Sensitivity, Recall and Specificity. The formulas to derive this measure are given in Eqs. (4.15), (4.16), (4.17).

$$\text{Precision} = \frac{\text{TP}}{\text{TP} + \text{FP}} \tag{4.15}$$

$$\text{Recall} = \frac{\text{TP}}{\text{TP} + \text{FN}} \tag{4.16}$$

$$\text{Specificity} = \frac{\text{TN}}{\text{TN} + \text{FP}} \tag{4.17}$$

The precision and recall can be combined together to calculate *F-measure*. A constant β controls the trade-off between precision and recall. Formula for calculating *F-measure* is given in Eq. (4.18).

$$\text{F} - \text{measure} = \frac{(\beta^2 + 1) * \text{Precision} * \text{Recall}}{\beta^2 * \text{Precision} + \text{Recall}} \tag{4.18}$$

In general β value is taken as 0.9 for better analysis of F-measure.

The most important question in any classification task is that *how precise is the accuracy rate estimate*. The accuracy rate estimate is more accurate based on the larger size of test set. This can be calculated by calculating the confidence interval for a given level of statistical significance. When we measure the accuracy on a test set, we are actually performing random experiments on different independent test sets. Let TS is the total test set and ITS is any independent test set from total test set. Let Acc_{TS} is the accuracy of the total test set and Acc_{ITS} is the accuracy of independent test set. Then the accuracy of classifier on total test set can be represented according to Eq. (4.19).

$$Acc_{TS} = Acc_{ITS} \pm Z_{CL} * SD_{ITS} \tag{4.19}$$

Where Z_{CL} is the value of a standard normal random variable associated with a desired confidence level CL. SD_{ITS} is the standard deviation of accuracy estimate Acc_{ITS}. Values of Z_{CL} for confidence level 90%, 95%, 98%, 99% are given in Table 4.5 assuming two-sided confidence intervals.

Standard deviation can be calculated as given in Eq. (4.20).

$$SD_{ITS} = \sqrt{(Acc_{ITS} * (1 - Acc_{ITS}))/n} \tag{4.20}$$

Table 4.5 Value of Z_{CL} on different confidence intervals

Confidence level	90%	95%	98%	99%
Value of Z_{CL}	1.64	1.96	2.33	2.58

where n is the number of data instances in any independent test set. Also, the validation of the results has been performed by using k-fold cross validation. Here, k value is chosen as 10. So the whole dataset is divided into 10 unique subsets. In each cycle of classification process one set is taken for testing purpose and rest of the sets are taken for training purpose. So total 10 cycles for classification task have been performed and the performance metrics have been calculated. Then the average of these metrics is taken as the final performance results. There is a very minute difference between the best performance results and average performance results through cross validation.

4.5.3 Results and Analysis

This section provides the detailed experimental evaluation results performed according to above said procedures. Different experiments have been performed to analyze the efficiency of classification process trained with IPSO for classifying epileptic seizure from nonepileptic ones. To prove this, it has been compared with other existing techniques such as RBFNN trained with gradient descent approach and RBFNN trained with conventional PSO algorithm. Figs. 4.4–4.6 show the graph for MSE with respect to the number of iterations for different datasets with different training algorithms. (Fig. 4.4 for RBFNN with GD approach, Fig. 4.5 for RBFNN with conventional PSO, and Fig. 4.6 for RBFNN with IPSO).

These graphs show the comparison of different methods. Tables 4.6 and 4.7 show the training and testing accuracy of different training methods. Table 4.6 shows different accuracies validated at a confidence level 95% and Table 4.7 shows accuracies validated at confidence level 98%.

Figs. 4.7 and 4.8 show the graphical representation for comparison of different training algorithms on four different datasets. Here, SET1 represents EEG data for epilepsy with set A and E. SET2 represents EEG data for epilepsy with set D and E. SET3 represents EEG data for epilepsy with set A, D, and E. SET4 represents EEG data for eye state prediction.

As per the previous discussions, other than accuracy several different measures have been considered for comparing different techniques. Those include precision, recall, specificity, and F-measure. Table 4.8 presents the values of these measures for RBFNN trained with GD approach. Table 4.9 presents the values of these measures for RBFNN trained with conventional PSO algorithm. Table 4.10 presents the values of these measures for RBFNN trained with IPSO algorithm. Similarly, the graphical

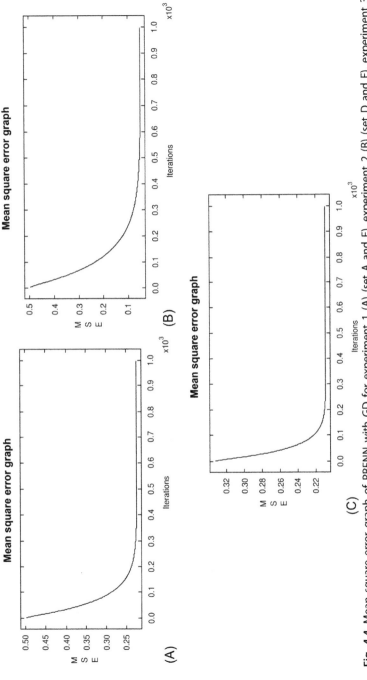

Fig. 4.4 Mean square error graph of RBFNN with GD for experiment 1 (A) (set A and E), experiment 2 (B) (set D and E), experiment 3 (C) (set A+D and E).

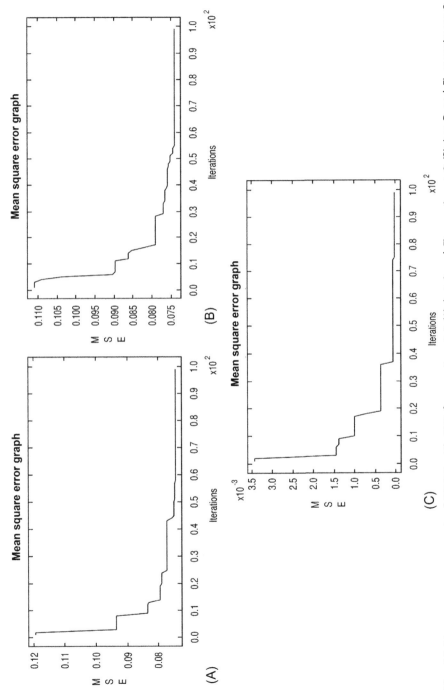

Fig. 4.5 Mean square error graph of RBFNN with PSO for experiment 1 (A) (set A and E), experiment 2 (B) (set D and E), experiment 3 (C) (set A+D and E).

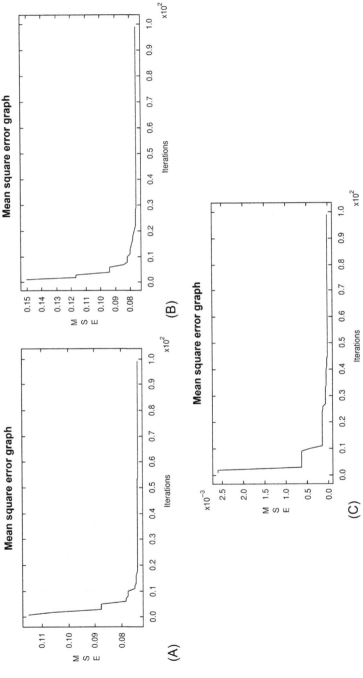

Fig. 4.6 Mean square error graph of RBFNN with IPSO for experiment 1 (A) (set A and E), experiment 2 (B) (set D and E), experiment 3 (C) (set A+D and E).

Table 4.6 Training and testing accuracy comparison of RBFNN (inverse multiquadric) trained with GD, PSO, and IPSO (confidence level 95%)

Datasets used in experiment	RBF trained with GD		RBF trained with general PSO		RBF trained with IPSO	
	Training accuracy	Testing accuracy	Training accuracy	Testing accuracy	Training accuracy	Testing accuracy
EEG for epilepsy (set A and E)	97.0 ± 0.033	70.0 ± 0.089	97.0 ± 0.033	96.00 ± 0.038	99.0 ± 0.019	99.0 ± 0.019
EEG for epilepsy (set D and E)	89.2 ± 0.060	84.0 ± 0.071	98.0 ± 0.027	96.00 ± 0.038	99.0 ± 0.019	97.0 ± 0.033
EEG for epilepsy (set A+D and E)	80.4 ± 0.063	75.3 ± 0.069	85.7 ± 0.056	78.6 ± 0.065	90.6 ± 0.046	84.6 ± 0.057
EEG for eye state prediction	90.6 ± 0.006	86.4 ± 0.007	93.4 ± 0.005	87.54 ± 0.0074	98.3 ± 0.0029	95.19 ± 0.0048

Table 4.7 Comparison of performance of RBFNN (Gaussian) trained with GD, PSO, and IPSO (confidence level 98%)

Datasets used in experiment	RBF trained with GD		RBF trained with general PSO		RBF trained with IPSO	
	Training accuracy	Testing accuracy	Training accuracy	Testing accuracy	Training accuracy	Testing accuracy
EEG for epilepsy (set A and E)	97.0 ± 0.039	70.0 ± 0.106	97.0 ± 0.039	96.00 ± 0.0456	99.0 ± 0.023	99.0 ± 0.023
EEG for epilepsy (set D and E)	89.2 ± 0.072	84.0 ± 0.085	98.0 ± 0.032	96.00 ± 0.0456	99.0 ± 0.023	97.0 ± 0.039
EEG for epilepsy (set A+D and E)	80.4 ± 0.075	75.3 ± 0.100	85.7 ± 0.066	78.6 ± 0.095	90.6 ± 0.055	84.6 ± 0.084
EEG for eye state prediction	90.6 ± 0.0078	86.4 ± 0.009	93.4 ± 0.006	87.54 ± 0.007	98.3 ± 0.003	95.19 ± 0.0057

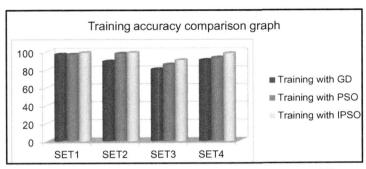

Fig. 4.7 Comparison of training accuracy for different datasets using different training techniques.

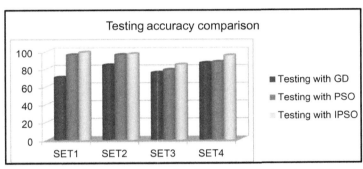

Fig. 4.8 Comparison of testing accuracy for different datasets using different training techniques.

Table 4.8 Other performance measures for RBFNN trained with GD approach

Datasets used in experiment	Precision	Recall	Specificity	F-measure
EEG for epilepsy (set A and E)	0.9	0.642	0.833	0.763
EEG for epilepsy (set D and E)	0.98	0.845	0.976	0.914
EEG for epilepsy (set A + D and E)	0.26	1.0	0.729	0.388
EEG for eye state prediction	1.0	0.782	1.0	0.889

Table 4.9 Other performance measures for RBFNN trained with PSO approach

Datasets used in experiment	Precision	Recall	Specificity	F-measure
EEG for epilepsy (set A and E)	0.92	1.0	0.926	0.954
EEG for epilepsy (set D and E)	0.92	1.0	0.926	0.954
EEG for epilepsy (set A + D and E)	0.7	1.0	0.869	0.808
EEG for eye state prediction	1.0	0.782	1.0	0.889

Table 4.10 Other performance measures for RBFNN trained with IPSO approach

Datasets used in experiment	Precision	Recall	Specificity	F-measure
EEG for epilepsy (set A and E)	0.98	1.0	0.98	0.988
EEG for epilepsy (set D and E)	0.94	1.0	0.943	0.966
EEG for epilepsy (set A + D and E)	0.78	1.0	0.9	0.865
EEG for eye state prediction	0.892	1.0	0.92	0.937

representations for comparing different techniques have been provided in Figs. 4.9–4.12, respectively. From these detailed experimental evaluation it can be strongly proved that RBFNN trained with IPSO algorithm outperforms other techniques for the classification of EEG signal in epileptic seizure identification.

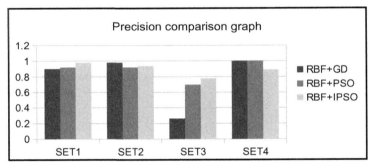

Fig. 4.9 Comparison of precision for different datasets using different training techniques.

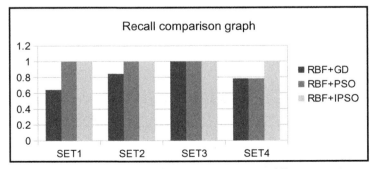

Fig. 4.10 Comparison of recall for different datasets using different training techniques.

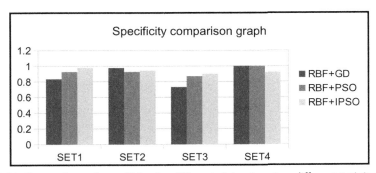

Fig. 4.11 Comparison of specificity for different datasets using different training techniques.

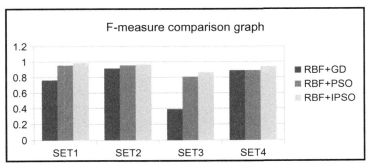

Fig. 4.12 Comparison of F-measure for different datasets using different training techniques.

4.6 SUMMARY

In this research work, a new IPSO algorithm has been used to train the RBFNN more efficiently to classify epileptic seizures. Also, this technique has been tested on another dataset, that is, Eye state prediction. This proposed technique has been compared with other available techniques (gradient decent, convention PSO) by rigorous and thorough practical implementations and experimental results. Hence, it is proved that the proposed technique outperforms other existing techniques. In this research domain, DWT is only used for analysis and statistical feature extraction from EEG datasets for epilepsy. For eye state prediction the dataset is already in the format for classification. In the next following chapter, an ABC optimized RBFNN classifier is proposed for the identification of epileptic seizure.

CHAPTER 5

ABC Optimized RBFNN for Classification of EEG Signal for Epileptic Seizure Identification

The brain signals usually generate certain electrical signals that can be recorded and analyzed for detection in several brain disorder diseases. These small signals are expressly called as EEG signals. This research work analyzes the epileptic disorder in human brain through EEG signal analysis by integrating the best attributes of ABC and RBFNNs. DWT technique was used for the extraction of the potential features from the signal. For classification of these signals, we have trained the RBFNN by a modified version of ABC algorithm. In the modified ABC, the onlooker bees are selected based on binary tournament unlike roulette wheel selection (RWS) of ABC. Additionally, kernels like Gaussian, multiquadric, and inverse multiquadric are used for measuring the effectiveness of the method in numerous mixtures of healthy segments, seizure-free segments, and seizure segments. Our experimental outcomes confirm that RBFNN with inverse multiquadric kernel trained with modified ABC is significantly better than RBFNNs with other kernels trained by ABC and modified ABC. In the previous chapter, for the classification of EEG signal, we have used RBFNN trained to optimize mean square error (MSE) by using a modified PSO algorithm. The main idea behind the selection of ABC over PSO technique is that, in ABC there is a less number of dependable parameters as compared to PSO. The effectiveness of this procedure has been verified by an experimental study on a benchmark dataset publicly available. The result of our experimental study revealed that the improvement of our attempt was significant and remarkable over RBF trained by gradient descent and our own modified PSO.

5.1 RELATED WORK

ABC optimization algorithm is a very fresh and new technique proposed in 2005. Hence, it can be stated that very less number of work has been

EEG Brain Signal Classification for Epileptic Seizure Disorder Detection
https://doi.org/10.1016/B978-0-12-817426-5.00005-3

performed using this technique as compared to PSO technique. But still lot of valuable and efficient research has been performed by different researchers using this ABC algorithm. Patrícia et al. [176] have proposed an ABC algorithm base clustering method to create a RBFNN classifier. They have suggested a new version of ABC that is called as cOptBees along with a heuristic model to automatically select different parameters of basis function that is to be used in RBFNN. They had verified the techniques in terms of decision boundaries generated and also accuracy of the classifier. They had also made a valid comparison of proposed technique with k-means, random center selection methods, and so on. Mustaffa et al. [177] have proposed least square support vector machine (LSSVM) using improved ABC for gasoline price forecasting. They have optimized the parameters of LSSVM using an improved version of ABC algorithm which is based on the conventional mutation operation used in GA. They have considered the fitness function as the mean absolute percentage error and root mean square percentage error. Karaboga and Ozturk [178] who is the inventor of ABC algorithm have proposed a novel clustering approach using ABC algorithm. They made an important invention regarding the task of clustering using the techniques of ABC to create homogeneous group of objects. They also compared their approach with other existing techniques such as PSO and other techniques from the literature. Nieto et al. [179] have used the ABC algorithm to design wavelet kernel-based SVM to predict cyanotoxin content in Trasona reservoir. Alshamlana et al. [180] have proposed a new gene selection technique for microarray data based on genetic bee colony algorithm. Yu and Duan [181] have used the ABC algorithm for information granulation in a fuzzy RBFNN in image fusion task. Garro et al. [182] have performed the classification of DNA microarray data using ANN and ABC algorithm.

5.2 ARTIFICIAL BEE COLONY ALGORITHM

ABC is a SI technique developed by Dervis Karaboga in 2005. Its main aim is to optimize numerical problems. It has been motivated from the foraging behavior of the honey bees. The basic nature or intelligence of a honey bee can be used for solving many real-life problems. Honey bees are one of the interesting swarms in nature. They have the skills like photographic memories, space-age sensory, and navigation systems. Honey bees are social insects that live in colonies (as shown in Fig. 5.1).

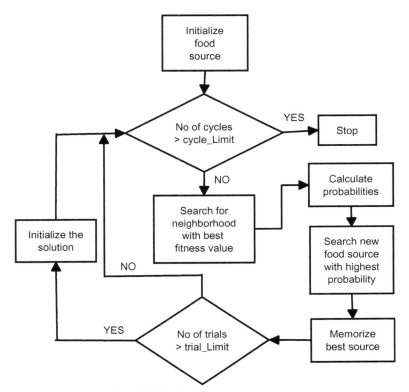

Fig. 5.1 Working procedure of ABC algorithm.

5.2.1 Architecture

In ABC algorithm there are mainly two types of bees, such as employed and unemployed bees. Unemployed bees can be again categorized as onlooker bees and scout bees. The initializations of food sources are done using the following formula in Eq. (5.1).

$$x_{mi} = l_i + \text{rand}(0, 1) * (u_i - l_i) \tag{5.1}$$

The employed bees search for a new food source having more nectar within the neighborhood of the food source in their memory. The neighbor food sources can be selected by using the following formula (as shown in Eq. 5.2).

$$v_{mi} = x_{mi} + \varphi_{mi}(x_{mi} - x_{ki}) \tag{5.2}$$

where m is the number of solutions, i, k are the number of parameters to optimize, and φ_{mi} is a random number. After selecting the neighborhoods, their fitness can be calculated using a fitness function. The fitness value of a solution can be calculated as follows (as shown in Eq. 5.3).

$$fit_m\left(\vec{x}_m\right) = \begin{cases} \dfrac{1}{1+f_m\left(\vec{x}_m\right)}, \text{ if } f_m\left(\vec{x}_m\right) \geq 0 \\ 1 + abs\left(f_m\left(\vec{x}_m\right)\right), \text{ if } f_m\left(\vec{x}_m\right) < 0 \end{cases} \qquad (5.3)$$

The bees that are waiting in the dancing area for taking decision on selecting a food source are called as onlooker bees. They select a food source depending on the probability of fitness values provided by employed bees according to the following formula (as shown in Eq. 5.4).

$$P_i = \frac{fit_i\left(\vec{x}_i\right)}{\displaystyle\sum_{i=1}^{FS} fit_i\left(\vec{x}_i\right)} \qquad (5.4)$$

Onlooker bees visit the food source that they select and identify a nearby modified source. They evaluate and choose between the original and new source. The employed bees whose sources were abandoned become scouts and go in search of new food sources. The scout discovers a new food source by employing Eq. (5.2), where *rand* is a random number between 0 and 1. The algorithm avoids getting into local optimum by having the scouts perform a random global search for new food sources.

5.2.2 Algorithm

Here, the algorithm for basic ABC was discussed that was suggested by its inventors. It generally pointed out the different essential phases of the procedure to implement ABC algorithm. It is described in Algorithm 5.1.

Algorithm 5.1. ABC Algorithm

Initialization Phase
REPEAT
    Employed Bees Phase
    Onlooker Bees Phase
    Scout Bees Phase
    Memorize the best solution achieved so far
UNTIL(Cycle = Maximum Cycle Number)

5.3 RBFNN WITH IMPROVED ABC ALGORITHM

This research work mainly focuses on classifying epileptic seizure patients from nonseizure patients by suitably trained RBFNNs. The trained RBFNN is developed by combining the best attributes of gradient descent trained RBFNN and modified ABC. The same architecture was discussed in Chapter 4 in Section 4.2.1 that was also used for this work.

5.3.1 Architecture of the Proposed Model

Initially, we adopt gradient dscent approach to train the RBFNN and then the trained parameters like centers, spreads, weights are feeding as the seed points of the ABC and modified ABC. The optimized parameters set up the final architecture of RBFNN to assign a class label to sample with no class label. The detailed flowchart of the proposed model is given in Fig. 5.2.

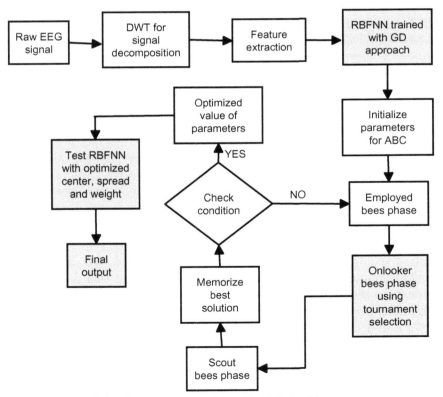

Fig. 5.2 Model of classification using RBFNN with ABC algorithm.

After the raw EEG signal is collected from the source, first of all it should be analyzed to discover the hidden characteristics or features of these signals. This can be performed using DWT technique, which decomposes the signal into several levels, thus extracts different statistical features. The EEG signal comes with 5 different sets (A, B, C, D, and E). Therefore the experimental work is divided into three parts. First part consists of A and E, the second consists set D and E, and the third is a collection of A and D with E. These datasets are now ready for classification work. In this work, we have taken three prominent kernels (discussed in Section 4.2) for the nodes of the hidden layer of RBFNN. These three kernels play the pivotal role in addition to the novel training algorithms while classifying these EEG signals.

By considering each individual kernel, RBFNN has been trained with gradient descent approach and then successively trained with the ABC. Then these intermediate values of different parameters of RBFNN will be used to initialize the solution vector for ABC algorithm. By using this algorithm the optimal values of center, width, and weight will be calculated. Here, the objective function is taken as the MSE as given in Eq. (5.5).

$$\text{MSE}(c, \sigma, w) = \frac{1}{n}\left(\sum_{i=1}^{n}\left(d_i - \sum_{j=1}^{m}w_j * H_j(X)\right)\right)^2 \tag{5.5}$$

With an objective of minimizing the error, ABC starts initializing the solutions (consists of three parameters, such as center, width, and weight) and then repeat the loop with required steps up to several runs/the limit, for which parameters will be optimized.

The ABC algorithm will proceed in three different phases: *Employed bees*, *Onlooker bees*, and *Scout bees* phase. To make the selection of onlooker bees easy and competitive, we replace the RWC mechanism by binary tournament. The inspiration of adopting this mechanism in ABC came from selection mechanism of GAs, in which randomly selected pair of bees will compete among each other to be selected depending on their fitness value.

Some of the comparisons made between RWS and tournament selection (TS) have been discussed later and diagrammatically shown in Fig. 5.3. RWS is difficult to solve real-world problems as it is suitable for only maximization problems. Fitness values have to be converted for solving minimization problems. RWS has no parameter to control selection pressure. TS preserves diversity, as it gives chance to all individuals to be selected. The detailed pseudo-code of proposed method is given in Algorithm 5.2.

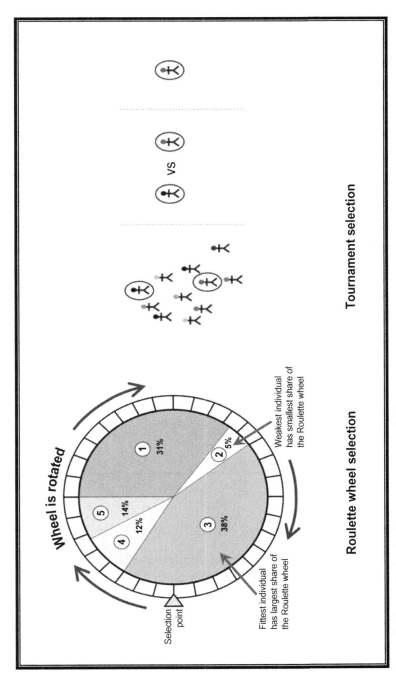

Fig. 5.3 Roulette wheel selection vs tournament selection.

5.3.2 Algorithm for the Proposed Model

Here, we presented the complete algorithm for our proposed technique of classification of epileptic seizures using RBFNN trained with modified ABC algorithm.

Algorithm 5.2. Algorithm for Proposed Model of RBF Trained With Modified ABC

Input: Preprocessed EEG dataset for epileptic seizure Identification.

Output: Class label prediction

Step1: Initialize and setup the parameters for RBFNN

Step2: Load the training sample for RBFNN and train the network

Step3: Initialize and setup parameters for ABC algorithm with Gradient Descent learning approach up to certain number of iterations.

Step4: Initialize solution (Center, Width, and Weight) for each food source in ABC algorithm to values of Center, width, and weight values found in step 2.

Step5: Find the fitness value using Equation 12 with the objective function defined in Eq. (5.5).

Step6: Set cycle = 0 and for each employed bee set trials = 0

Step7: Repeat Step 8 to 15 until cycle < CycleLimit

   Step8: For each employed bee find the neighbor bee by using Eq. (5.2)

   Step9: Find the fitness value and make a greedy selection.

   Step10: If the solution is selected set trials = 0, Otherwise trial++

   Step11: Select an onlooker bee using tournament selection

   Step12: Find the fitness value and make a greedy selection.

   Step13: If the solution is selected set trials = 0, Otherwise trial++

   Step14: Memorize the best solution found so far

   Step15: If for any solution trials = TrialLimit

      Step16: Abandon the solution (Scout bee)

      Step17: Randomly initialize the solution and go to step 7.

Step18: Set the RBF network by taking the optimized parameters (Center, Width, and Weight) found in above steps.

Step19: Test the network with EEG test samples.

5.4 EXPERIMENTAL STUDY

This section describes the detailed analysis of the experimental work that is carried out for our proposed model. The computational complexity of the proposed algorithm may vary for different datasets depending on its size. The parametric values may vary accordingly.

5.4.1 Dataset Preparation and Environment

For this study, five sets of EEG signals for Epileptic seizure identification have been collected from publicly available source [37]. There are three combinations of these sets taken for experimental study, that is, set A and E (experiment 1), set D and E (experiment 2), and set A + D and E (experiment 3). For DWT of EEG signal, we have used the MATLAB toolbox for wavelet transform. After this, all other programming codes for entire experimental work have been designed using Java platform (JDK 1.8 with Eclipse Mars IDE).

5.4.2 Parameters

All the three datasets are first taken for classification using RBFNN with *gradient descent learning algorithm*. This algorithm is evaluated by taking different values of learning parameter (η) from 0.5 to 1.0. By different experimental evaluation we found that for EEG dataset the Gaussian and inverse multiquadric basis functions outperform as compared to multiquadric function. After that a deep research has been done to enhance the performance of RBFNN using ABC algorithm. For ABC algorithm there are several parameters that have been set initially as follows:

a. Colony size = 40 (i.e., number of employed bees + onlooker bees)
b. No. of food sources = 20 (colony size/2)
c. maxLimit = 100 (number of times a food source can be improved)
d. maxCycle = 50 (number of cycles for foraging)
e. Number of parameters to optimize = 880 (number of center parameters + number of spread parameters + number of weight parameters)
f. $lb = -1$, $ub = +1$ (*lb*—lower bound and *ub*—upper bound for parameters)
g. Fitness function, $f(c, \sigma, w) = \frac{1}{n}\left(\sum_{i=1}^{n} d_j - \sum_{j=1}^{m} w_j * H_j(x)\right)^2$, where for Gaussian function $H_j(x)$ is given in Eq. (4.2), multiquadric function $H_j(x)$ is given in Eq. (4.3) and inverse multiquadric function $H_j(x)$ is given in Eq. (4.4) in Section 4.2.

For RBFNN there are mainly three types of basis functions that are used such as Gaussian, multiquadric, and inverse multiquadric. But due to the high performance of Gaussian and inverse multiquadric, the multiquadric function has been ignored. Classification results of the classifiers were collected by a confusion matrix. In a confusion matrix, each cell contains the number of exemplars classified for the corresponding combination of desired and actual network outputs. The test performance of the methods was determined by the computation of the following statistical parameters for different experiments.

Experiment 1 (set A and E):

$$\text{Specificity} = \frac{EE}{EE + AE} \tag{5.6}$$

$$\text{Sensitivity} = \frac{AA}{AA + EA} \tag{5.7}$$

The accuracy of the model is defined as:

$$\text{Accuracy} = \frac{AA + EE}{EE + AA + EA + AE} \tag{5.8}$$

where AA is the count of cases that belong to the A class and are predicted as A (true positives); AE is the count of cases that belong to the E class and are predicted as A (false positives); EE is the count of cases that belong to the E class and are predicted as E (true negatives); and EA is the count of cases that belong to the A class and are predicted as E (false negatives). Similarly, the meaning of A and E is defined as follows: A is the EEG signals recorded from healthy volunteers with eyes open, E is the EEG signals recorded from epilepsy patients during epileptic seizures.

Similarly, the performance metrics of other experiments have been defined like Eqs. (5.6)–(5.8). However, the notations are different. The results have been validated by k-fold cross validation. Here, k value is chosen as 10. So, the whole dataset is divided into 10 unique subsets, that is, in each cycle of classification process, one set is taken for testing purpose and rest of the sets are taken for training purpose. As a result, total 10 cycles for classification task have been performed and the performance metrics was computed. Hence, the averages of these metrics were taken as the final performance results. It was observed that there was a very minute difference between the best performance results and average performance results obtained through cross validation.

5.4.3 Result and Analysis

Figs. 5.4–5.6 show the MSE graph for RBFNN with (A) Gaussian RBF and (B) inverse multiquadric RBF with varying learning parameter for Experiment number 1, 2, and 3, respectively. From these experiments, it has been concluded that for inverse multiquadric RBF there is no effect of the learning parameter. For Gaussian RBF as the value of learning parameter increases, the MSE quickly tends to its minima and for certain value of learning parameter it gives minimum MSE.

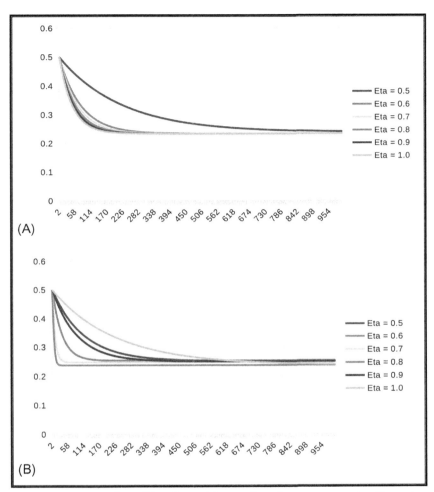

Fig. 5.4 Experiment 1 (A and E) MSE graph for gradient descent approach with varying η value: (A) Gaussian RBF and (B) inverse multiquadric RBF.

Now, RBFNN has been trained using ABC algorithm and for performance evaluation the MSE graph has been plotted for different runs of ABC. Figs. 5.7 and 5.8 show the variation in MSE for 50 runs of ABC algorithm with (A) Gaussian RBF and (B) inverse multiquadric RBF. It is being clearly observed that using ABC training algorithm the performance of RBFNN with inverse multiquadric function has been enhanced. The MSE has been successfully reduced to 0.07 (approximately) after training the network with ABC optimization algorithm.

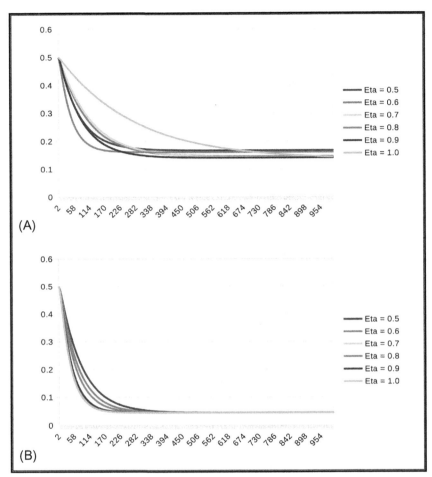

Fig. 5.5 Experiment 2 (D and E) MSE graph for gradient descent approach with varying η value: (A) Gaussian RBF and (B) inverse multiquadric RBF.

Table 5.1 presents the comparison of *sensitivity*, *specificity*, and *accuracy* between two training approaches, gradient descent and ABC of RBFNN. Clearly, it shows that the performance of RBFNN trained with ABC algorithm is improved over traditional gradient descent approach in all three experiments. Table 5.2 presents the comparison of performance of general ABC and our modified ABC. These results were taken from the best performances of the classifiers. Tables 5.3 and 5.4 present the results taken from

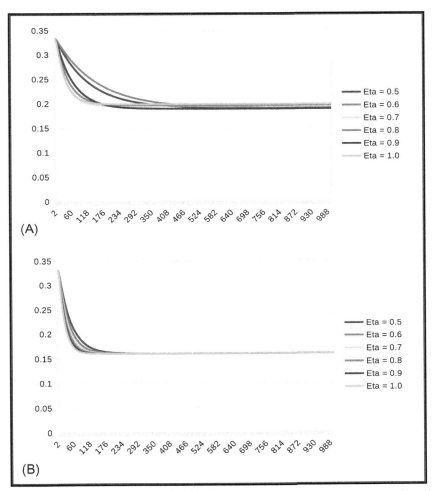

Fig. 5.6 Experiment 3 (A+D and E) MSE graph for gradient descent approach with varying η value: (A) Gaussian RBF and (B) inverse multiquadric RBF.

the 10-fold cross validated classifiers. Evidently, there is an extremely not as much of difference in these results. There is somehow an improvement in the performance for experiment 1 and 3, but there is a huge improvement in experiment 3. From this experimental evaluation, it is clearly proved that modified ABC algorithm can classify EEG data for epilepsy identification with highest accuracy.

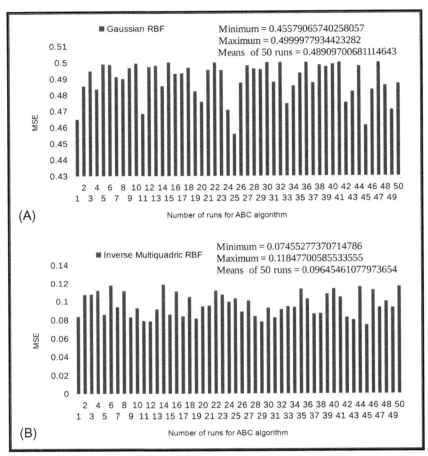

Fig. 5.7 Experiment 1 (set A and E) MSE graph for ABC trained RBF: (A) Gaussian RBF and (B) inverse multiquadric RBF.

5.4.4 Performance Comparison Between Modified PSO and Modified ABC Algorithm

According to the research done in Chapter 4 and this chapter in this section we have made a performance comparison between RBFNN trained with improved PSO and modified ABC algorithm. Table 5.5 describes the accuracy comparison between the two proposed techniques. From this comparison it is clearly understood that the modified ABC algorithm outperforms as compared to the improved PSO algorithm. Also, some other advantages of ABC over PSO include less number of dependable parameters in ABC as compared to PSO. Hence, the processing speed and value approximation is more in ABC algorithm.

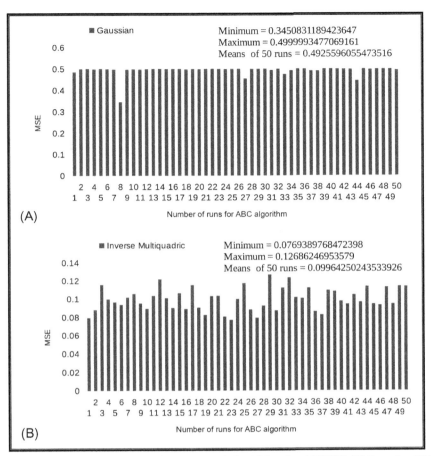

Fig. 5.8 Experiment 2 (set D and E) MSE graph for ABC trained RBFNN: (A) Gaussian RBF and (B) inverse multiquadric RBF.

Table 5.1 Performance comparison between GD learning and ABC learning with inverse multiquadric RBFNN

Experiments	Specificity		Sensitivity		Accuracy	
	RBFNN with GD	RBFNN with ABC	RBFNN with GD	RBFNN with ABC	RBFNN with GD	RBFNN with ABC
Set A and E	84.1	95.7	65.7	93.4	71.5	92.5
Set D and E	100.0	100.0	80.0	88.5	87.5	93.5
Set A + D and E	76.8	95.6	63.2	66.0	73.3	81.6

Table 5.2 Performance comparison between ABC learning and modified ABC learning with inverse multiquadric RBFNN

	Specificity		Sensitivity		Accuracy	
Experiments	RBFNN with ABC	RBFNN with MABC	RBFNN with ABC	RBFNN with MABC	RBFNN with ABC	RBFNN with MABC
Set A and E	95.7	100.0	93.4	98.0	92.5	99.5
Set D and E	100.0	100.0	88.5	97.0	93.5	98.5
Set A + D and E	95.6	96.0	66.0	68.4	81.6	83.0

Table 5.3 Performance comparison between GD learning and ABC learning with inverse multiquadric RBFNN with 10-fold cross validation

	Average specificity		Average sensitivity		Average accuracy	
Experiments	RBFNN with ABC	RBFNN with GD	RBFNN with ABC	RBFNN with ABC	RBFNN with GD	RBFNN with ABC
Set A and E	84.1	95.7	65.7	93.4	71.5	91.5
Set D and E	100.0	100.0	80.0	88.5	87.5	93.5
Set A + D and E	76.8	95.6	63.2	66.0	73.3	81.6

Table 5.4 Performance comparison between ABC learning and modified ABC learning with inverse multiquadric RBFNN with 10-fold cross validation

	Average specificity		Average sensitivity		Average accuracy	
Experiments	RBFNN with ABC	RBFNN with MABC	RBFNN with ABC	RBFNN with MABC	RBFNN with ABC	RBFNN with MABC
Set A and E	95.7	100.0	93.4	98.6	91.5	99.0
Set D and E	100.0	100.0	88.5	95.4	94.5	97.5
Set A + D and E	95.6	96.0	66.0	68.4	80.6	82.3

Table 5.5 Performance comparison between modified ABC trained RBFNN and improved PSO trained RBFNN

	Accuracy	
Experiments	RBFNN with MABC	RBFNN with IPSO
Set A and E	99.5	99.0
Set D and E	98.5	97.0
Set A + D and E	83.0	84.6

5.5 SUMMARY

Our approach is primarily based on ABC algorithm, which is a new and robust algorithm used for training the RBFNN. However, we noticed that after adopting the binary TS in the onlooker bees' phase, our novel approach for classifying epileptic seizure vs nonepileptic seizure in three distinct experiments has been performing significantly better than gradient descent and ABC trained RBFNN. The performance of ABC trained RBFNN algorithm is compared with gradient descent approach trained RBFNN which is mostly used by the researchers. Finally, our concluding remark says ABC can be applied successfully to enhance the performance of RBF network for classification of EEG signal for detecting epileptic seizures. Moreover, the preprocessing of EEG signal is done by using DWT which is also an important requirement before doing the classification. In the next chapter, we discuss the overall details of the book and some of the future aspects that need to be attended.

CHAPTER 6

Conclusion and Future Research

This research work is set out to investigate the invention and role of EEG signal analysis and classification for epileptic seizure identification. EEG signals are the brain waves generated in the human brain that signifies the different tasks performed by a person. Depending on the state of signals generated in the psyche, the persons react differently to a position. Hence, careful and strict analysis of these signals can solve a set of diseases occurring in our mind. Straight off the days there are very highly equipped machineries available to read these signs from our learning ability. But after recording sometimes it is really difficult for the medical specialists to diagnose and conclude any specific neurological disease. Hence, there are plentiful opportunities for computer scientists to produce a computer-based model that can distinguish a specific neurological disease after proper analysis of these signals. In this book, we have presented some of our contributions in this area. This chapter describes in brief about our findings and constraints for this research study along with some scopes for future insights.

6.1 FINDINGS AND CONSTRAINTS

Along with the introduction about this book in Chapter 1, we have represented our contributions in different chapters. In Chapter 2, an extensive literature survey addressing different facets of the whole study has been performed. These aspects may include EEG signal analysis methods, where different techniques proposed by researchers for analyzing the EEG signal are addressed. This analysis is normally different from general signal analysis techniques because these signs are very nonstationary in nature. The different analysis techniques may include DWT, entropy measures, Lyapunov exponents, and so on. Another aspect of the study is preprocessing of the EEG signals after recording from brain to remove any kind of noise or artifacts. These artifacts are nothing but some unwanted contents in EEG signal which may happen due to any outside noises. The researchers have also proposed different techniques to get rid of these artifacts completely and then break down the signal for more dependable functioning. After that different tasks have been identified that can be performed by EEG signals. Generally, this work is purely based on analysis and classification of EEG signal for epilepsy identification. But beside this there are several other tasks that can be performed by an EEG signal such as detecting sleep disorder, eye state prediction and designing BCI, and so on. Then a survey of different methods used for classification of EEG signal was discussed that included a comparison between machine learning-based classifiers and statistical modeling-based classifiers. After that a survey of different machine learning techniques applied for detection of epileptic seizures were introduced. In Chapter 3 an empirical survey of different machine learning techniques applied for designing classifier model for epilepsy identification was performed. The different techniques used are MLPNN with different learning algorithms (such as BP, RPROP, and MUR), SVM with different kernel functions (such as linear, polynomial, and RBF kernel), PNN, RNN, and RBFNN. From this survey, it was concluded that the RBFNN is the simplest network used, compared to others. Besides the training speed is lot more eminent. But the performance of this technique is not up to the grade. Hence, a thought was given and it was decided to enhance the performance of this network by implementing some swarm-based optimization techniques as learning algorithm. In Chapter 4, a novel hybrid technique to train RBFNN by optimizing the parameters of this network using the PSO algorithm was offered. PSO is one of the most effective optimization technique used so far. But again, the existing PSO algorithm was modified to enhance the performance by adding a new technique for finding the value of the inertia weight used in the calculation of velocity update. The experimental evaluations have proved that the performance of the proposed technique is higher than RBFNN trained with GD approach and as well as RBFNN trained with the conventional PSO algorithm. In Chapter 5 we switched from PSO to ABC algorithm because of the large number of dependable parameters in PSO. We also suggested a change to the existing ABC algorithm, where TS in place of RWS in onlookers bees phase was used. Besides, the operation of the proposed technique has been shown to be more efficient by different experimental evaluations.

One of the major constraints in this research work that was faced is the accumulation of the real-time data from different patients. This involves depth knowledge of machinery expertise it was genuinely difficult to gather real-time data from medicals. For this constraint a publicly available data for research in epilepsy identification was considered. This constraint would be addressed by us in our future study.

6.2 FUTURE RESEARCH WORK

With these concluding remarks our future research work is yet to produce more efficient and robust classifier model for epilepsy identification and also for other tasks of EEG. Some of the future works are stated as follows:

- Designing of more robust signal analysis and feature extraction technique.
- Addressing the issue of class imbalance problem in epilepsy identification.
- Including other various machine learning techniques for further enhancing the empirical study.
- Addressing the problem of multiclass classification for epileptic seizure identification.

References

[1] Niedermeyer E, Lopes da Silva F. Electroencephalography: basic principles. In: Clinical applications and related fields. New York: Lippincot Williams & Wilkins; 2004.

[2] Abou-Khalil B, Musilus KE. Atlas of EEG & seizure semiology. Philadelphia: Elsevier Publication; 2005.

[3] Bickford RG, Brimm J, Berger I. Application of compressed spectral array in clinical EEG. In: Kellaway P, Peterson I, editors. Automation of clinical electroencephalography. New York: Raven-Press; 1973. p. 55–64.

[4] Sanei S, Chambers JA. EEG signal processing. New York: Wiley Publications; 2013.

[5] Zhang Q, Lee M. Fuzzy-GIST for emotion recognition in natural scene images. In: International conference on development and learning; 2009. p. 1–7.

[6] Pfurtscheller G, Neuper C, Schlogl A, Lugger K. Separability of EEG signals recorded during right and left motor imagery using adaptive autoregressive parameters. IEEE Trans Rehabil Eng 1998;6(3):316–25.

[7] Teplan M. Fundamentals of EEG measurement. Meas Sci Rev 2002;2(2):1–11.

[8] Fisch B. Fisch and Spehlmann's EEG primer. 3rd ed. New York: Elsevier; 1999.

[9] Towle VL, Bolaños J, Suarez D, Tan K, Grzeszczuk R, Levin DN, Cakmur R, Frank SA, Spire JP. The spatial location of EEG electrodes: locating the best fitting sphere relative to cortical anatomy. Electroencephalogr Clin Neurophysiol 2003;86(1):1–6.

[10] Aurlien H, Gjerde IO, Aarseth JH, Karlsen B, Skeidsvoll H, Gilhus NE. EEG background activity described by a large computerized database. Clin Neurophysiol 2004;15(3):665–73.

[11] Rabbi AF, Rezai RF. A fuzzy logic system for seizure onset detection in intracranial EEG. Comput Intell Neurosci 2012;2012:1–13.

[12] Walter C, Cierniak G, Gerjets P, Rosenstiel W, Bogdan M. Classifying mental states with machine learning algorithms using alpha activty decline. In: European symposium on artificial neural networks, computational intelligence and machine learning; 2011. p. 405–10.

[13] Wu X, Kumar V, Quinlan JR, Ghosh J, Yang Q, Motoda H, McLachlan GJ, Ng A, Liu B, Yu PS, Zhou ZH, Steinbach M, Hand DJ, Steinberg D. Top 10 algorithms in data mining. Knowl Inf Syst 2007;14(1):1–37.

[14] Siuly YL, Wen P. Clustering technique-based least square support vector machine for EEG signal classification. Comput Methods Programs Biomed 2008;104(3):358–72.

[15] Jahankhani P, Kodogiannis V, Revett K. EEG signal classification using wavelet feature extraction and neural networks. In: Proceedings of IEEE John Vincent Atanasoff international symposium on modern computing; 2006. p. 120–4.

[16] Guler ID. Adaptive neuro-fuzzy inference system for classification of EEG signals using wavelet coefficients. J Neurosci Methods 2005;148:113–21.

[17] Riedmiller M, Braun H. A direct adaptive method for faster back propagation learning: the RPROP algorithm. In: Proceedings of IEEE international conference on neural networks, vol. 1. 1993. p. 586–91.

[18] Guler NF, Ubeyli ED, Guler I. Recurrent neural networks employing Lyapunov exponents for EEG signals classification. Expert Syst Appl 2005;29(3):506–14.

[19] Petrosian A, Prokhorov D, Homan R, Dascheiff R, Wunsch D. Recurrent neural network based prediction of epileptic seizures in intra- and extracranial EEG. Neurocomputing 2000;30(1–4):201–18.

[20] Petrosian A, Prokhorov DV, Lajara-Nanson W, Schiffer RB. Recurrent neural network-based approach for early recognition of Alzheimer's disease in EEG. Clin Neurophysiol 2001;112(8):1378–87.

[21] Übeyli ED. Analysis of EEG signals by implementing eigen vector methods/recurrent neural networks. Digital Signal Process 2009;19(1):134–43.

[22] Derya E. Recurrent neural networks employing Lyapunov exponents for analysis of ECG signals. Expert Syst Appl 2010;37(2):1192–9.

[23] Saad EW, Prokhorov DV, Wunsch DC. Comparative study of stock trend prediction using time delay, recurrent and probabilistic neural networks. IEEE Trans Neural Netw 1998;9(6):1456–70.

[24] Lima CAM, Coelho ALV, Eisencraft M. Tackling EEG signal classification with least squares support vector machines: a sensitivity analysis study. Comput Biol Med 2010;40(8):705–14.

[25] Limaa CAM, Coelhob ALV. Kernel machines for epilepsy diagnosis via EEG signal classification: a comparative study. Artif Intell Med 2011;53(2):83–95.

[26] Siuly YL, Wen P. Classification of EEG signals using sampling techniques and least square support vector machines. Rough Sets Knowl Technol 2009;5589:375–82.

[27] Übeyli ED. Least squares support vector machine employing model-based methods coefficients for analysis of EEG signals. Expert Syst Appl 2010;37(1):233–9.

[28] Dhiman R, Saini JS. Priyanka, genetic algorithms tuned expert model for detection of epileptic seizures from EEG signatures. Appl Soft Comput 2014;19:8–17.

[29] Xu Q, Zhou H, Wang Y, Huang J. Fuzzy support vector machine for classification of EEG signals using wavelet-based features. Med Eng Phys 2009;31(7):858–65.

[30] Yuan Q, Zhou W, Li S, Cai D. Epileptic EEG classification based on extreme learning machine and nonlinear features. Epilepsy Res 2011;96(1–2):29–38.

[31] Liu Y, Zhou W, Yuan Q, Chen S. Automatic seizure detection using wavelet transform and SVM in long-term intracranial EEG. IEEE Trans Neural Syst Rehabil Eng 2012;20(6):749–55.

[32] Shoeb A, Kharbouch A, Soegaard J, Schachter S, Guttag J. A machine learning algorithm for detecting seizure termination in scalp EEG. Epilepsy Behav 2011;22 (1):36–43.

[33] Bonyadi MR, Michalewicz Z. Particle swarm optimization for single objective continuous space problems: a review. Evol Comput 2017;25(1):1–54.

[34] Trelea IC. The particle swarm optimization algorithm: convergence analysis and parameter selection. Inf Process Lett 2003;85(6):317–25.

[35] Bratton D, Blackwell T. A simplified recombinant PSO. J Artif Evol Appl 2008; (3):1–10.

[36] Meissner M, Schmuker M, Schneider G. Optimized particle swarm optimization and its application to artificial neural network training. BMC Bioinform 2006;7:125–35.

[37] EEG Data. [Online], http://www.meb.unibonn.de/science/physik/eegdata.html; 2001.

[38] Andrzejak RG, Lehnertz K, Mormann F, Rieke C, David P, Elger CE. Indications of nonlinear deterministic and finite-dimensional structures in time series of brain electrical activity: dependence on recording region and brain state. Phys Rev E 2001;64 (6):1–6.

[39] Gotman J. Automatic recognition of epileptic seizure in the EEG. Electroencephalogr Clin Neurophysiol 1982;54(5):530–40.

[40] Subasi A, Alkan A, Kolukaya E, Kiymik MK. Wavelet neural network classification of EEG signals by using AR model with MLE pre-processing. Neural Netw 2005;18 (7):985–97.

[41] Subasi A, Gursoy MI. EEG signal classification using PCA, ICA, LDA and support vector machines. Expert Syst Appl 2010;37(12):8659–66.

[42] Petrosian A. Kolmogorov complexity of finite sequences and recognition of different preictal EEG patterns. In: Proceedings of the 8th IEEE symposium on computer-based medical systems; 1995. p. 212–7.

[43] Higuchi T. Approach to an irregular time series on the basis of the fractal theory. Physica D 1988;31(2):277–83.

[44] Balli T, Palaniappan R. A combined linear & nonlinear approach for classification of epileptic EEG signals. In: Proceedings of the 4th international IEEE/EMBS conference on neural engineering (NER '09); 2009. p. 714–7.

[45] Inouye T, Shinosaki K, Sakamoto H, Toi S, Ukai S, Iyama A, Katsuda Y, Hirano M. Quantification of EEG irregularity by use of the entropy of the power spectrum. Electroencephalogr Clin Neurophysiol 1991;79(3):204–10.

[46] Pincus SM, Gladstone IM, Ehrenkranz RA. A regularity statistic for medical data analysis. J Clin Monit Comput 1991;7(4):335–45.

[47] Roberts SJ, Penny W, Rezek I. Temporal and spatial complexity measures for electroencephalogram based brain-computer interfacing. Med Biol Eng Comput 1999;37(1):93–8.

[48] Peng C-K, Havlin S, Stanley HE, Goldberger AL. Quantification of scaling exponents and crossover phenomena in non stationary heartbeat time series. Chaos 1995;5(1):82–7.

[49] Hjorth B. EEG analysis based on time domain properties. Electroencephalogr Clin Neurophysiol 1970;29(3):306–10.

[50] Gandhi T, Panigrahi BK, Anand S. A comparative study of wavelet families for EEG signal classification. Neurocomputing 2011;74:3051–7.

[51] Li Y, Luo M, Li K. A multiwavelet-based time-varying model identification approach for time–frequency analysis of EEG signals. Neurocomputing 2016;193:106–14.

[52] Guo L, Rivero D, Dorado J, Munteanu CR, Pazos A. Automatic feature extraction using genetic programming: an application to epileptic EEG classification. Expert Syst Appl 2011;38:10425–36.

[53] Raymer M, Punch W, Goodman E, Kuhn L. Genetic programming for improved data mining: application to the biochemistry of protein interactions. In: Proceedings of the first annual conference on genetic programming; 1996. p. 375–80.

[54] Sherrah J. Automatic feature extraction for pattern recognition [PhD thesis]. South Australia: The University of Adelaide; 1998.

[55] Tzallas A, Tsipouras M, Fotiadis D. A time-frequency based method for the detection of epileptic seizures in EEG recordings. In: The proceedings of 20th IEEE international symposium on computer-based medical systems; 2007. p. 135–40.

[56] Acharya UR, Molinari F, Sree SV, Chattopadhyay S, Hoong K, Surig JS. Automated diagnosis of epileptic EEG using entropies. Biomed Signal Process Control 2012;7:401–8.

[57] Nasehi S, Pourghassem H. A new feature dimensionally reduction approach based on general tensor discriminant analysis in EEG signal classification. In: The proc. of international conference on intelligent computation and bio-medical instrumentation; 2011. p. 188–91.

[58] Oveisi F. EEG signal classification using nonlinear independent component analysis. In: The proceeedings of IEEE international conference on acoustics, speech and signal processing; 2009. p. 361–4.

[59] Delgado Saa JF, Gutierrez MS. EEG signal classification using power spectral features and linear discriminant analysis: a brain computer Interface application. In: The proceedings of eighth Latin American and Caribbean conference for engineering and technology, innovation and development for the Americas; 2010. p. 1–7.

[60] Subha DP, Joseph PK, Acharya R, Lim CM. EEG signal analysis: a survey. J Med Syst 2010;34:195–212.

[61] Pradhan N, Dutt DN. Data compression by linear prediction for storage and transmission of EEG signals. Int J Biomed Comput 1994;35(3):207–17.

[62] Tzyy-Ping J, Makeig S, Mckeown MJ, Bell AJ, Te-Won L, Sejnowski TJ. Imaging brain dynamics using independent component analysis. Proc IEEE 2001;89(7):1107–22.

[63] Welch PD. The use of fast Fourier transform for the estimation of power spectra: a method based on time averaging over short, modified periodograms. IEEE Trans Audio Electroacoust 1967;AU-15(2):70–3.

[64] Cheng M, Jia W, Gao X, Gao S, Yang F. Mu rhythm-based cursor control: an offline analysis. Clin Neurophysiol 2004;115(4):745–51.

[65] Kubler A, Nijboer F, Mellinger J, Vaughan TM, Pawelzik H, Schalk G, McFarland DJ, Birbaumer N, Wolpaw JR. Patients with ALS can use sensorimotor rhythms to operate a brain–computer interface. Neurology 2005;64(10):1775–7.

[66] Fabiani GE, McFarland DJ, Wolpaw JR, Pfurtscheller G. Conversion of EEG activity into cursor movement by a brain–computer interface. IEEE Trans Neural Syst Rehabil Eng 2004;12(3):331–8.

[67] Babiloni F, Cincotti F, Lazzarini L, Millan J, Mourino J, Varsta M, Heikkonen J, Bianchi L, Marciani MG. Linear classification of low-resolution EEG patterns produced by imagined hand movements. IEEE Trans Rehabil Eng 2000;8(2):186–8.

[68] Babiloni F, Cincotti F, Bianchi L, Pirri G, Millan JR, Mourino J, Salinari S, Marciani MG. Recognition of imagined hand movements with low resolution surface Laplacian and linear classifiers. Med Eng Phys 2001;23(5):323–8.

[69] Cincotti F, Bianchi L, Millan J, Mourino J, Salinari S, Marciani MG, Babiloni F. Brain–computer interface: the use of low resolution surface Laplacian and linear classifiers for the recognition of imagined hand movements. In: The proceedings of 23rd annual international conference of the IEEE engineering in medicine and biology society; 2001. p. 665–8.

[70] Cincotti F, Mattia D, Babiloni C, Carducci F, Salinari S, Bianchi L, Marciani MG, Babiloni F. The use of EEG modifications due to motor imagery for brain–computer interfaces. IEEE Trans Neural Syst Rehabil Eng 2003;11(2):131–3.

[71] Qin L, Deng J, Ding L, He B. Motor imagery classification by means of source analysis methods. In: The proceedings of the 26th annual international confernece of the IEEE engineering in medicine and biology society; 2004. p. 4356–8.

[72] Muller KR, Curio G, Blankertz B, Dornhege G. Combining features for BCI. In: Advances in neural information processing systems, vol. 15. Cambridge, Mass: The MIT Press; 2003. p. 1115–22.

[73] Millan JR, Mourino J, Marciani MG, Babiloni F, Topani F, Canale I, Heikkonen J, Kaski K. Adaptive brain interfaces for physically-disabled people. In: The proceedings of the 20th annual international conference of the IEEE engineering in medicine and biology society; 1998. p. 2008–11.

[74] Schalk G, Wolpaw JR, McFarland DJ, Pfurtscheller G. EEG-based communication: presence of an error potential. Clin Neurophysiol 2000;111(12):2138–44.

[75] Bayliss JD, Ballard DH. Single trial P300 recognition in a virtual environment. In: The proceedings of the international ICSC symposium on soft computing in biomedicine; 1999. Technical Report.

[76] Erfanian A, Erfani A. ICA-based classification scheme for EEG-based brain–computer interface: the role of mental practice and concentration skills. In: The proceedings of the 26th annual international conference of the IEEE engineering in medicine and biology society; 2004. p. 235–8.

[77] Gao X, Xu N, Hong B, Gao S, Yang F. Optimal selection of independent components for event-related brain electrical potential enhancement. In: The proc. of the IEEE international workshop on biomedical circuits and systems; 2004. pp. S3/5/INV–S3/5/1-4.

[78] Peterson DA, Knight JN, Kirby MJ, Anderson CW, Thaut MH. Feature selection and blind source separation in an EEG-based brain–computer interface. EURASIP J Appl Signal Process 2005;(19):3128–40.

[79] Wu RC, Liang SF, Lin CT, Hsu CF. Applications of event-related-potential-based brain–computer interface to intelligent transportation systems. In: The proceedings of the IEEE international conference on networking, sensing and control; 2004. p. 813–8.

[80] Serby H, Yom-Tov E, Inbar GF. An improved P300-based brain–computer interface. IEEE Trans Neural Syst Rehabil Eng 2005;13(1):89–98.

[81] Xu N, Gao X, Hong B, Miao X, Gao S, Yang F. BCI competition 2003–data set IIb: enhancing P300 wave detection using ICA-based subspace projections for BCI applications. IEEE Trans Biomed Eng 2004;51(6):1067–72.

[82] Chapin JK, Moxon KA, Markowitz RS, Nicolelis MA. Real-time control of a robot arm using simultaneously recorded neurons in the motor cortex. Nat Neurosci 1999;2(7):664–70.

[83] Guan C, Thulasidas M, Wu J. High performance P300 speller for brain–computer interface. In: The proceedings of IEEE international workshop on biomedical circuits and systems; 2004. pp. S3/5/INV-S3/13–16.

[84] Hu J, Si J, Olson BP, He J. Principle component feature detector for motor cortical control. In: The proceedings of 26th annual international conference of the IEEE engineering in medicine and biology society; 2004. p. 4021–4.

[85] Isaacs RE, Weber DJ, Schwartz AB. Work toward real-time control of a cortical neural prothesis. IEEE Trans Rehabil Eng 2000;8(2):196–8.

[86] Lee H, Choi S. PCA-based linear dynamical systems for multichannel EEG classification. In: The proceedings of 9th international conference on neural information processing; 2002. p. 745–9.

[87] Yoon H, Yang K, Shahabi C. Feature subset selection and feature ranking for multivariate time series. IEEE Trans Knowl Data Eng 2005;17(9):1186–98.

[88] Li Y, Cichocki A, Guan C, Qin J. Sparse factorization pre processing-based offline analysis for a cursor control experiment. In: The proceedings of IEEE international workshop on biomedical circuits and systems; 2004. pp. S3/5/INV-S3/5/5–8.

[89] Trejo LJ, Wheeler KR, Jorgensen CC, Rosipal R, Clanton ST, Matthews B, Hibbs AD, Matthews R, Krupka M. Multimodal neuroelectric interface development. IEEE Trans Neural Syst Rehabil Eng 2003;11(2):199–204.

[90] Lemm S, Blankertz B, Curio G, Muller KR. Spatio-spectral filters for improving the classification of single trial EEG. IEEE Trans Biomed Eng 2005;52(9):1541–8.

[91] Bashashati A, Ward RK, Birch GE. A new design of the asynchronous brain–computer interface using the knowledge of the path of features. In: The proceedings of 2nd IEEE-EMBS conference on neural engineering; 2005. p. 101–4.

[92] Borisoff JF, Mason SG, Bashashati A, Birch GE. Brain–computer interface design for asynchronous control applications: improvements to the LF-ASD asynchronous brain switch. IEEE Trans Biomed Eng 2004;51(6):985–92.

[93] Fatourechi M, Bashashati A, Borisoff JF, Birch GE, Ward RK. Improving the performance of the LF-ASD brain–computer interface by means of genetic algorithm. In: The proceedings of the IEEE symposium on signal processing and information technology; 2004. p. 38–41.

[94] Yu Z, Mason SG, Birch GE. Enhancing the performance of the LF-ASD brain–computer interface. In: The proceedings of the 24th annual international conference of the IEEE engineering in medicine and biology society; 2002. p. 2443–4.

[95] Peters BO, Pfurtscheller G, Flyvbjerg H. Automatic differentiation of multichannel EEG signals. IEEE Trans Biomed Eng 2001;48(1):111–6.

[96] Makeig S, Bell A, Jung T, Sejnowski T. Independent component analysis of electroencephalographic data. Adv Neural Inf Proces Syst 1995;8:145–51.

[97]   Clark DC, Gonzalez RC. Optimal solution of linear inequalities with application to pattern recognition. IEEE Trans Pattern Anal Mach Intell 1981;3(6):643–55.

[98]   Li Y, Gao X, Liu H, Gao S. Classification of single-trial electroencephalogram during finger movement. IEEE Trans Biomed Eng 2004;51(6):1019–25.

[99]   Liu HS, Gao X, Yang F, Gao S. Imagined hand movement identification based on spatio-temporal pattern recognition of EEG. In: The proceedings of the IEEE-EMBS conference on neural engineering; 2003. p. 599–602.

[100] Wang Y, Zhang Z, Li Y, Gao X, Gao S, Yang F. BCI competition 2003–data set IV: an algorithm based on CSSD and FDA for classifying single-trial EEG. IEEE Trans Biomed Eng 2004;51(6):1081–6.

[101] Vidal JJ. Real-time detection of brain events in EEG. Proc IEEE 1977;65(5):633–41.

[102] Zibulevsky M, Pearlmutter BA. Blind source separation by sparse decomposition. Neural Comput 2001;13(4):863–82.

[103] Pregenzer M, Flotzinger DM, Pfurtschellar G. Automated feature selection with a distinction sensitive learning vector quantizer. Neurocomputing 1996;11:19–29.

[104] Obeid I, Wolf PD. Evaluation of spike-detection algorithms for a brain-machine interface application. IEEE Trans Biomed Eng 2004;51(6):905–11.

[105] Sanchez JC, Carmena JM, Lebedev MA, Nicolelis MA, Harris JG, Principe JC. Ascertaining the importance of neurons to develop better brain-machine interfaces. IEEE Trans Biomed Eng 2004;51(6):943–53.

[106] Zamir ZR. Detection of epileptic seizure in EEG signals using linear least squares preprocessing. Comput Methods Prog Biomed 2016;133:95–109.

[107] Boashash B, Ouelha S. Automatic signal abnormality detection using time-frequency features and machine learning: a newborn EEG seizure case study. Knowl-Based Syst 2016;106:38–50.

[108] He L, Liu B, Hu D, Wen Y, Wan M, Long J. Motor imagery EEG signals analysis based on Bayesian network with Gaussian distribution. Neurocomputing 2016;188:217–24.

[109] Ellenrieder N, Dana J, Frauscher B, Gotman J. Sparse asynchronous cortical generators can produce measurable scalp EEG signals. NeuroImage 2016;138:123–33.

[110] Banerjee A, Sanyal S, Patranabis A, Banerjee K, Guhathakurta T, Sengupta R, Ghosh D, Ghose P. Study on brain dynamics by non linear analysis of music induced EEG signals. Physica A 2016;444:110–20.

[111] Das AB, Bhuiyan MIB. Discrimination and classification of focal and non-focal EEG signals using entropy-based features in the EMD-DWT domain. Biomed Signal Process Control 2016;29:11–21.

[112] Fu K, Qu J, Chai Y, Zou T. Hilbert marginal spectrum analysis for automatic seizure detection in EEG signals. Biomed Signal Process Control 2015;18:179–85.

[113] Cuellar M, Harkrider AW, Jenson D, Thornton D, Bowers A, Saltuklaroglu T. Time–frequency analysis of the EEG mu rhythm as a measure of sensorimotor integration in the later stages of swallowing. Clin Neurophysiol 2016;127:2625–35.

[114] Gao D, Li M, Li J, Liu Z, Yao D, Li G, Liu T. Effects of various typical electrodes and electrode gels combinations on MRI signal-to-noise ratio and safety issues in EEG-fMRI recording. Biocybern Biomed Eng 2016;36(1):9–18.

[115] Peker M. An efficient sleep scoring system based on EEG signal using complex-valued machine learning algorithms. Neurocomputing 2016;207:165–77.

[116] Wang J, Yang C, Wang R, Yu H, Cao Y, Liu J. Functional brain networks in Alzheimer's disease: EEG analysis based on limited penetrable visibility graph and phase space method. Physica A 2016;460:174–87.

[117] Orhan U, Hekim M, Ozer M. EEG signals classification using the K-means clustering and a multilayer perceptron neural network model. Expert Syst Appl 2011;38 (10):13475–81.

[118] Tangkraingkij P, Lursinsap C, Sanguansintukul S, Desudchit T. Selecting relevant EEG signal locations for personal identification problem using ICA and neural network. In: The proceedings of the eighth IEEE/ACIS international conference on computer and information science; 2009. p. 616–21.

[119] Kousarrizi MRN, Ghanbari AA. Feature extraction and classification of EEG signals using wavelet transform, SVM and artificial neural networks for brain computer interfaces. In: The proceedings. of the international joint conference on bioinformatics, systems biology and intelligent computing; 2009. p. 352–5.

[120] Murugesan M, Sukanesh R. Automated detection of brain tumor in EEG signals using artificial neural networks. In: The proceedings of the international conference on advances in computing, control, and telecommunication technologies; 2009. p. 284–8.

[121] Jia H. Neural network in the application of EEG signal classification method. In: The proceedings of the seventh international conference on computational intelligence and security; 2011. p. 1325–7.

[122] Alzoubi O, Koprinska I, Calvo RA. Classification of brain-computer interface data. In: The proceedings of the seventh Australasian data mining conference 87; 2008.

[123] Skinner BT, Nguyen HT, Liu DK. Classification of EEG signals using a genetic-based machine learning classifier. In: The proceedings of the IEEE engineering in medicine and biology society; 2007. p. 3120–3.

[124] Liang N, Saratchandran P, Huang G, Sundararajan N. Classification of mental tasks from EEG signals using extreme learning machine. Int J Neural Syst 2006;16(1):29–38.

[125] Ioannides AA, Poghosyan V, Dammers J, Streit M. Real-time neural activity and connectivity in healthy indiviuals and schizophrenia patients. NeuroImage 2004;23 (2):1–2.

[126] Garrett D, Peterson DA, Anderson CW, Thaut MH. Comparison of linear and nonlinear methods for EEG signal classification. IEEE Trans Neural Syst Rehabil Eng 2011;11(2):141–4.

[127] Aris S, Taib N, Lias S, Norizam S. Feature extraction of EEG signals and classification using FCM. In: The proceedings of the UK sim 13th international conference on modelling and simulation; 2011. p. 54–8.

[128] Selim AE, Wahed MA, Kadah YM. Machine learning methodologies in brain-computer interface systems. In: The proceedings of the Cairo International Biomedical Engineering conference; 2008. p. 1–5.

[129] Guler I, Ubeyli ED. Multiclass support vector machines for EEG-signals classification. IEEE Trans Inf Technol Biomed 2007;11(2):117–27.

[130] Tomasevic NM, Neskovic AM, Neskovic NJ. Artificial neural network based approach to EEG signal simulation. Int J Neural Syst 2012;22(3):1–12.

[131] Lee H, Choi S. PCA+HMM+SVM for EEG pattern classification. Signal Process Appl 2003;1:541–4.

[132] Garrett D, Peterson DA, Anderson CW, Thaut MH. Comparison of linear, nonlinear, and feature selection methods for EEG signal classification. IEEE Trans Neural Syst Rehabil Eng 2003;11(2):141–4.

[133] Guo L, Rivero D, Dorado J, Rabunal JR, Pazos A. Automatic epileptic seizure detection in EEGs based on line length feature and artificial neural networks. J Neurosci Methods 2010;191:101–9.

[134] Nigam V, Graupe D. A neural-network-based detection of epilepsy. Neurol Res 2004;26(1):55–60.

[135] Srinivasan V, Eswaran C, Sriraam N. Artificial neural network based epileptic detection using time-domain and frequency-domain features. J Med Syst 2005;29 (6):647–60.

[136] Kannathal N, Acharya U, Lim C, Sadasivan P. Characterization of EEG—a comparative study. Comput Methods Prog Biomed 2005;80(1):17–23.

[137] Kannathal N, Choo M, Acharya U, Sadasivan P. Entropies for detection of epilepsy in EEG. Comput Methods Prog Biomed 2005;80(3):187–94.

[138] Polat K, Günes S. Classification of epileptiform EEG using a hybrid system based on decision tree classifier and fast Fourier transform. Appl Math Comput 2007;187 (2):1017–26.

[139] Subasi A. EEG signal classification using wavelet feature extraction and a mixture of expert model. Expert Syst Appl 2007;32(4):1084–93.

[140] Ocak H. Automatic detection of epileptic seizures in EEG using discrete wavelet transform and approximate entropy. Expert Syst Appl 2009;36(2):2027–36.

[141] Yuan Q, Zhou W, Yuan S, Li X, Wang J, Jia G. Epileptic EEG classification based on kernel sparse representation. Int J Neural Syst 2014;24(4):1–13.

[142] Anusha KS, Mathews MT, Puthankattil SD. Classification of normal and epileptic EEG signal using time & frequency domain features through artificial neural network. In: The proceedings of the international conference on advances in computing and communications; 2012. p. 98–101.

[143] Prince PGK, Hemamalini R. A survey on soft computing techniques in epileptic seizure detection. In: The proceedings of the emerging trends in robotics and communication technologies; 2010. p. 377–80.

[144] Subasi A, Ercelebi E. Classification of EEG signals using neural network and logistic regression. Comput Methods Prog Biomed 2005;78(2):87–99.

[145] Mirowski PW, LeCun Y, Madhavan D, Kuzniecky R. Comparing SVM and convolutional networks for epileptic seizure prediction from intracranial EEG. In: The proceedings of the IEEE workshop on machine learning for signal processing; 2008. p. 244–9.

[146] Quirago RQ, Schurmann M. Functions and sources of event-related EEG alpha oscillations studied with the wavelet transform. Clin Neurophysiol 1999;110(4):643–54.

[147] Mirowski P, Madhavan D, LeCun Y, Kuzniecky R. Classification of patterns of EEG synchronization for seizure prediction. Clin Neurophysiol 2009;120(11):1927–40.

[148] Majumdar K. Human scalp EEG processing: various soft computing approaches. Appl Soft Comput 2011;11(8):4433–47.

[149] Teixeira CA, Direito B, Bandarabadi M, Quyen MLV, Valderrama M, Schelter B, Schulze-Bonhage A, Navarro V, Sales F, Dourado A. Epileptic seizure predictors based on computational intelligence techniques: a comparative study with 278 patients. Comput Methods Prog Biomed 2014;114(3):324–36.

[150] Siuly and Y. Li. A novel statistical algorithm for multiclass EEG signal classification. Eng Appl Artif Intell 2014;34:154–67.

[151] Siuly YL, Wen P. Analysis and classification of EEG signals using a hybrid clustering technique. In: The proceedings of the IEEE/ICME international conference on complex medical engineering; 2010. p. 34–9.

[152] Saastamoinen A, Pietila T, Varri A, Lehtokangas M, Saarinen J. Waveform detection with RBF network application to automated EEG analysis. Neurocomputing 1998;20 (1–3):1–13.

[153] Derya E. Lyapunov exponents/probabilistic neural networks for analysis of EEG signals. Expert Syst Appl 2010;37(2):985–92.

[154] Gupta L, McAvoy M, Phegley J. Classification of temporal sequences via prediction using the simple recurrent neural network. Pattern Recogn 2010;33(10):1759–70.

[155] Gupta L, McAvoy M. Investigating the prediction capabilities of the simple recurrent neural network on real temporal sequences. Pattern Recogn 2000;33(12):2075–81.

[156] Cristianini N, Taylor JS. Support vector and kernel machines. London: Cambridge University Press; 2001.

[157] Taylor JS, Cristianini N. Support vector machines and other kernel-based learning methods. London: Cambridge University Press; 2000.

[158] Vatankhaha M, Asadpourb V, FazelRezaic R. Perceptual pain classification using ANFIS adapted RBF kernel support vector machine for therapeutic usage. Appl Soft Comput 2013;13(5):2537–46.

[159] Heaton J. Programming neural networks with Encog 3 in Java. In: Heaton research. 2nd ed. USA: Heaton Research, Inc.; 2011.

[160] Finley KH, Dynes JB. Electroencephalographic studies in epilepsy: a critical analysis. Brain 1942;65(1):256–65.

[161] Lehnertz K. Non-linear time series analysis of intracranial EEG recordings in patients with epilepsy—an overview. Int J Psychophysiol 1999;34(1):45–52.

[162] Alicata FM, Stefanini C, Elia M, Ferri R, Del Gracco S, Musumeci SA. Chaotic behavior of EEG slow-wave activity during sleep. Electroencephalogr Clin Neurophysiol 1996;99(6):539–43.

[163] Acharya UR, Sree SV, Swapna G, Joy Martis R, Suri JS. Automated EEG analysis of epilepsy: a review. Knowl-Based Syst 2013;45:147–65.

[164] Satapathy SK, Jagadev AK, Dehuri S. An empirical analysis of training algorithms of neural networks: a case study of EEG signal classification using Java framework. In: Advances in intelligent systems and computing, vol. 309. Springer; 2015. p. 151–60.

[165] Fathi V, Montazer GA. An improvement in RBF learning algorithm based on PSO for real time applications. Neurocomputing 2013;111:169–76.

[166] Mazurowski MA, Habas PA, Zurada JM, Lo JY, Baker JA, Tourassi GD. Training neural network classifiers for medical decision making: the effects of imbalanced datasets on classification performance. Neural Netw 2008;21(2–3):427–36.

[167] Zhang J, Lok T, Lyu M. A hybrid particle swarm optimization-back-propagation algorithm for feed forward neural network training. Appl Math Comput 2007;185(2):1026–37.

[168] Ge HW, Qian F, Liang YC, Du WL, Wang L. Identification and control of nonlinear systems by a dissimilation particle swarm optimization-based elman neural network. Nonlinear Anal Real World Appl 2008;9(4):1345–60.

[169] Zhao L, Yang Y. PSO-based single multiplicative neuron model for time series prediction. Expert Syst Appl 2009;36:2805–12.

[170] Guerra FA, Coelho LDS. Multi-step ahead nonlinear identification of lorenz's chaotic system using radial basis neural network with learning by clustering and particle swarm optimization. Chaos 2008;35(5):967–79.

[171] Dehuri S, Roy R, Cho SB, Ghosh A. An improved swarm optimized functional link artificial neural network (ISO-FLANN) for classification. J Syst Softw 2012;85(6):1333–45.

[172] Dash CSK, Dash AP, Dehuri S, Cho SB, Wang GN. DE + RBFNs based classification: a special attention to removal of inconsistency and irrelevant features. Eng Appl Artif Intel 2013;26(10):2315–26.

[173] Dehuri S, Cho SB. Evolutionary optimized features in functional link neural network for classification. Expert Syst Appl 2010;37(6):4379–91.

[174] Qasem SN, Shamsuddin SM. Hybrid learning enhancement of RBF network based on particle swarm optimization. Found Comput Intell 2009;1:19–29.

[175] Shi Y, Eberhart RC. A modified particle swarm. In: Proceedings of the IEEE international conference on evolutionary computation; 1998. p. 1945–50.

[176] Patrícia D, Cruz F, Maia RD, da Silva LA, de Castro LN. Bee RBF: a bee-inspired data clustering approach to design RBF neural network classifiers. Neurocomputing 2016;172:427–37.

[177] Mustaffa Z, Yusof Y, Kamaruddin SS. Gasoline price forecasting: an application of LSSVM with improved ABC. Procedia Soc Behav Sci 2014;129:601–9.

[178] Karaboga D, Ozturk C. A novel clustering approach: artificial bee colony (ABC) algorithm. Appl Soft Comput 2011;11:652–7.

[179] P. J. G. Nieto, E. García-Gonzalo, J. R. A. Fernández, C. Díaz Muñiz, A hybrid wavelet kernel SVM-based method using artificial bee colony algorithm for predicting the cyanotox in content from experimental cyano bacteria concentrations in the Trasona reservoir, J Comput Appl Math, 2016 [in press].

[180] Alshamlana HM, Badra GH, Alohali YA. Genetic bee colony(GBC) algorithm: a new gene selection method for microarray cancer classification. Comput Biol Chem 2015;56:49–60.

[181] Yu J, Duan H. Artificial bee colony approach to information granulation-based fuzzy radial basis function neural networks for image fusion. Optik 2013;124(17):3103–11.

[182] Garro BA, Rodríguez K, Vázquez RA. Classification of DNA microarrays using artificial neural networks and ABC algorithm. Appl Soft Comput 2016;38:548–60.

INDEX

Note: Page numbers followed by *f* indicate figures and *t* indicate tables.